To my mentors:
CCC, DAH, CEH, DMV

CHIMPANZEE MATERIAL CULTURE

CHIMPANZEE MATERIAL CULTURE

Implications for Human Evolution

W. C. McGREW

Professor, Department of Sociology and Anthropology and Department of Zoology, Miami University

Winner of the 1996 W. W. Howells Book Prize of the American Anthropological Association

Published by the Press Syndicate of the University of Cambridge
The Pitt Building, Trumpington Street, Cambridge CB2 1RP
40 West 20th Street, New York, NY 10011–4211, USA
10 Stamford Road, Oakleigh, Melbourne 3166, Australia

First published 1992
Reprinted 1994, 1996

Printed in Great Britain at Bell & Bain Ltd, Glasgow

A catalogue record for this book is available from the British Library

Library of Congress cataloguing in publication data

McGrew, W.C. (William Clement), 1944–
Chimpanzee material culture: implications for human evolution /
W. C. McGrew
p. cm.
Includes bibliographical references (p.) and index.
ISBN 0 521 41303 6. – ISBN 0 521 42371 6 (paperback)
1. Chimpanzees – Behavior. 2. Tool use in animals. 3. Human
evolution. 4. Social evolution. 5. Material culture. 6. Hunting
and gathering societies. I. Title
QL737.P96M44 1992
599.88′440451–dc20 92–13551 CIP

ISBN 0 521 41303 6 hardback
ISBN 0 521 42371 6 paperback

Contents

Preface

I first looked at chimpanzees in 1972 at the Delta Regional Primate Research Center in Covington, Louisiana. Caroline Tutin and I had been sent there by David Hamburg and Stanford University to make ourselves useful while waiting for research clearance from Tanzania. As soon as permission was granted, we were to begin research in the Gombe National Park, under the direction of Jane Goodall. Meanwhile, at Delta we inherited from Emil Menzel two resources of great importance: a one-acre enclosure with seven chimpanzees and the patient guidance of Pal Midgett. As we stood amid the loblolly pines on that gloriously sunny January day and listened to the greeting hoots of Gigi and Co., it was the start of something good.

Nineteen years later, I have made 13 trips to Africa to study wild chimpanzees, lasting from 1 to 8 months and totalling over 4 years in the field. Five of these trips were to Tanzania, to work either at Gombe (1972, 1973, 1992) or Mahale (1974) or both (1982). Five (1976–1979) were to Senegal, to work at Mt. Assirik in the Parc National du Niokolo-Koba. Three were to Gabon, first to Belinga (1981), then to the Lopé Reserve (1985, 1990).

Many articles, chapters, notes, reviews, etc., have appeared since 1972, reporting the results of our studies, but until now no synthesis has been attempted. This volume aims to tie together a varied set of findings on tool-use and related activities. Of course, this is only a fraction of the field-work that has been done, but it is a start. Many of the chapters have been through several reincarnations, so that the originals may now be unrecognisable: Chapter 1's first version was given in 1985 in a symposium of the British Social Biology Council in London and published in their journal (McGrew, 1985); Chapter 3 combines two efforts, one given in a joint symposium of the British Ecological Society and the Royal Anthropological Institute in Durham (McGrew, 1989*c*) and another given to the Fondation Fyssen in Versailles (McGrew, 1992*a*); Chapter 4's earliest version was published in *Man* (McGrew & Tutin, 1978); Chapter 5 started as a contribution to a symposium of the Wenner-Gren Foundation at Burg Wartenstein (McGrew, 1979); Chapter 6 originated in a paper to the International Conference on Hunting and Gathering Societies in London and another version was published elsewhere (McGrew, 1987); Chapters 2 and 7–10 were purpose-written for this volume, although sharp-eyed readers may recognise lots of bits and pieces.

Every author has biases, so it seems advisable to set out as many of mine as possible. First, I am a naturalist, interested most in what happens in the real world of nature, rather than what can be made to happen in the artificial world of captivity. Thus, this volume is likely to be biassed towards field-studies instead of laboratory ones. Second, I am an empiricist, committed to data. The most elegant idea in the world is nothing more than that unless someone tests it. Thus, this monograph favours explicitly presented, rigorously analysed and statistically tested data, whenever possible, in preference to impressions, anecdotes or speculation. Third, I believe that science only works in the public domain, where findings are published and accessible to all. Thus, this volume tries to cite books and journals, and omits manuscripts, theses and personal communications, unless these are essential. Fourth, I am monolingual, so this effort is biassed toward English-language publications, for which I apologise. Finally, I am an evolutionist, having been imprinted as an undergraduate in the natural and not the social sciences. Perhaps this is a damning impediment for looking at culture, but I hope not.

Acknowledgements

Every bit of my field-work has been collaborative, and without such good colleagues in the bush, little would have got done. Two stand out: Caroline Tutin and Anthony Collins. They are simply the best, and my gratitude to them cannot be measured. Each of the others listed below knows what we shared and how grateful I am: Byron Alexander, Donna Anderson, Jim Anderson, Pamela Baldwin, Rugema Bambaganya, Adriano Bandora, Hassani Bituru, Stella Brewer, Peter Buirski, Curt Busse, Sue Chambers, Jean-Ives Collet, John Crocker, Michel Fernandez, Stephanie Hall, Stewart Halperin, Rebecca Ham, Paul

Harmatz, Mike Harrison, Sal Harrison, Carrie Hunter, Alimasi Kasulamemba, Awadhi Kasulamemba, Desider Kazon, Hank Klein, Mark Leighton, Petro Leo, Hamisi Matama, Hilali Matama, Norman McBeath, Peter Meic, Nancy Merrick, Hamisi Mkono, Juma Mkukwe, Kit Morris, Esilom Mpongo, Yahaya Ntabilio, Nancy Nicolson, Ramadhani Nyundo, Ann Pierce, Frans Plooij, Hetty Plooij, Anne Pusey, David Riss, Liz Rogers, Moshi Sadiki, Mohamedi Seifu, Kassim Selemani, Yasini Selemani, Martin Sharman, Joan Silk, Sara Simpson, Mitzi Thorndahl, Lee White, Jean Wickings, Liz Williamson, Richard Wrangham. Finally I am especially grateful to Jane Goodall for giving me more than one chance.

Equally important in the field are companions. For fresh eggs, Christmas treats, ice-cold beer, cathartic bops, but most of all, tolerance and steadfastness, I thank John and Eleanor Allen, Barbara Behrens, Faye Benedict, Blanche and Tony Brescia, Moshi Bunengwa, David Bygott, Robert Caputo, Ramji Dharsi, Gustavo Gandini, Jeanette Hanby, Junichiro Itani, Julie Johnson, Hugo van Lawick, Phyllis Lee, Claude Lucazeau, Cricket Lyman, Muriel MacKenzie, Patrice Marty, Jim Moore, Leanne and Mike Nash, Helen Neely, Lisa Nowell, Juliet Oliver, Nigel Orbell, Jon Pollock, Craig Packer, Rafaella Savinelli, Chuck de Sieyes, Michael Shaw, Hitomi and Yukio Takahata, Erasmus Tarimo, Yukimaru Sugiyama, Emilie van Zinnicq-Bergmann.

Generosity in science is no more clear than when unpublished data, photographs and manuscripts are shared. No one has been more constant in this over almost 20 years than Toshisada Nishida. I also thank Shigeru Azuma, Alison and Noel Badrian, Pamela Baldwin, Christophe Boesch, Stella Brewer, David Bygott, Jennifer Byrne, Richard Carroll, Anthony Collins, Mike Fay, Michel Fernandez, Jane Goodall, Dick Grove, Alison Hannah, Stewart Halperin, Bob Harding, John Hart, Barbara King, Adriaan Kortlandt, Jurgen Lethmate, Pal Midgett, Jim Moore, Jo van Orshoven, Wendell Oswalt, Vernon Reynolds, Jorge Sabater Pí, Leonn Satterthwait, Tom Struhsaker, Yukimaru Sugiyama, Akira Suzuki, Yukio Takahata, Caroline Tutin, Elisabetta Visalberghi, George Whitesides, Liz Williamson, Richard Wrangham, Tom Wynn.

Every research project is an expedition, and for sanity-saving letters, specialist supplies, sound advice, stimulating conversations, unnoticed references, loaned equipment, scrounged spares, and 1001 other things, I thank Phil Bock, Barry Bolton, Dick Byrne, Arnold Chamove, Eva

Crane, Sharon File, Gordon Gallup, Thomas Geissman, Bill Gotwald, Peter Gerone, Carol Gonzales, Cliff Henty, the late Helmut Hofer, Hilly Kaplan, Jane Lancaster, Bob Lavery, Robin Law, Chris Longhurst, Rainer Lorenz, Linda Marchant, Dorothy, Bill and Larry McGrew, Beth Merrick, Wendell Oswalt, Art Riopelle, Liz Rushton, John Russell, Anne Russon, the late Tom Tutin, Shigeo Uehara, Karen Valley, Gillian Watt, Andy Whiten, Adrienne Zihlman.

Many persons made key contributions, from clerical to conceptual, to the gestation of this volume. Some read drafts of chapters, but Rob Foley, Carol George, Sarah Hrdy, Tim Ingold and Toshisada Nishida read it all, for which I am most grateful. Alison Bowes, my PhD supervisor, was a paragon of patience and wisdom over more years than she ever anticipated. I also thank Pamela Baldwin, Anthony Collins, Mike Harrison, Hilly Kaplan, Jürgen Lethmate, Caroline Tutin and Tom Wynn for critical reading. Clerically, I was immensely helped by Cathie Francis, Anne Goldie, Kirsty Moore, Jocelyn Murgatroyd and Fay Somerville. I apologise to them all for errors remaining, for which I take sole responsibility.

Research is expensive, and the following bodies provided funds and equipment, sometimes crucially. I thank the American Philosophical Society, Boise Fund, Carnegie Trust for the Universities of Scotland, W.T. Grant Foundation, L.S.B. Leakey Foundation, L.S.B. Leakey Trust, Nuffield Foundation, Royal Zoological Society of Scotland, Science and Engineering Research Council, Stanford University, University of Stirling, Wenner-Gren Foundation for Anthropological Research.

Finally, because wild chimpanzees live in Africa, I have been a guest abroad, dependent on the hospitality and sufferance of the citizens, officials and agencies of other countries. In Gabon, I thank Centre International de Recherches Médicales de Franceville, L'Institut de Recherche sur l'Ecologie Tropicale. In Senegal, I thank Délégation Générale de Recherche Scientifique et Technique, Service des Parcs Nationaux. In Tanzania, I thank Tanzania Commission for Science and Technology, Serengeti Wildlife Research Institute, Tanzania National Parks, Japan International Cooperation Agency, Gombe Stream Wildlife Research Centre, Mahale Mountains Wildlife Research Centre.

Stirling W.C. McGrew
September 1991

1
Patterns of culture?

> The beasts of prey and finally the higher apes slowly came to rely upon other than biological adaptations, and upon the consequent increased plasticity the foundations were laid, bit by bit, for the development of intelligence.
>
> *Ruth Benedict (1935)*

Introduction

Imagine the following set of incidents:

1. A chimpanzee at Assirik repeatedly bashes the hard-shelled fruit of a baobab tree against one of its exposed roots. Eventually the fruit cracks open and the ape eats its contents. Earlier the chimpanzee ate the fruits of two other kinds of palm trees, but as no oil palms were available, these could not be eaten.
2. A chimpanzee at Gombe sits for an hour in the crown of an oil palm tree, patiently extracting the fruits. These are prised out one by one but processed by the mouthful: the fibrous outer husk is chewed to a wad and sucked dry, then both it and the undamaged nut inside are spat out or swallowed.
3. A chimpanzee at Kasoje walks through a grove of fruiting oil palms. Overhead in the trees, vervet monkeys consume the fruits while below bush pigs crunch the discarded nuts. The ape ignores the palms, though earlier on the same day several other domesticated plants used by the local humans, such as banana, mango and sugar cane, were eaten.
4. A chimpanzee at Lopé climbs an oil palm tree in a clearing in the rain forest. The oily husks are eagerly eaten, especially as this is the lean time of the dry season, but other equally nutritious kinds of nuts, also common and accessible, remain untouched.

Table 1.1. *Six key sites for understanding chimpanzees as predators on oil palms*

Site	Country	Sources
Assirik	Senegal	McGrew *et al.* (1988)
Gombe	Tanzania	Goodall (1968)
Kasoje	Tanzania	Nishida & Uehara (1983)
Lopé	Gabon	Tutin *et al.* (1991)
Tai	Ivory Coast	Boesch & Boesch (1990)
Bossou	Guinea	Sugiyama & Koman (1987)

5. A chimpanzee at Tai places a panda nut on a root anvil deep in the forest, then pounds the nut with a hammer of quartz. The hard shell shatters, revealing four almond-like kernels to be eaten. Hundreds of panda shells are strewn about, and both stone hammer and wooden anvil show signs of wear from prolonged re-use. Oil palms are abundant but the ape eats only their leaves and pith, not the fruits.
6. A chimpanzee at Bossou sits beneath an oil palm near a village, smashing open nut after nut. She uses one rock as a hammer and a larger one as an anvil. The working surfaces of both are pitted from constant use. Earlier the ape ate the outer husk, and now the fruit is being re-used, but this time it is the kernel that is extracted from the nut.

These are but six of many variations on a simple theme: the relationship between an animal predator, the chimpanzee, and a plant prey, the oil palm. The revealed variety illustrates an equally simple point: the subsistence of our closest living relation defies easy generalisation. More like ourselves, and less like other animals, chimpanzees show flexibility of action that is apt for an apparently unlimited range of contexts.

Before tackling the implications of the last statement and of this provocatively titled chapter, I will say a bit more about the prey, fill in the picture for the six sites introduced above (see Table 1.1), and add some findings from other places in Africa where chimpanzees and oil palms have been studied. Only then will I try to justify my theft of Benedict's (1935) phrase.

The prey

The oil palm is the most economically important of the palms in Africa. Its walnut-sized fruit is a drupe, the fibrous mesocarp of which provides copious amounts of oil for cooking or for industrial use, after only minimal processing. Within the husk is a hard-shelled nut containing an edible kernel. Besides being rich in energy and fatty acids, palm oil is a source of vitamin A (Hartley, 1966). The kernel is high in fats and protein, as well as elements such as calcium and phosphorus. Even the sap, often tapped and fermented by local people as 'palm wine', is nutritious. For the Efe pygmies of Zaïre, palm nuts provide 9% of total caloric intake, which is more than the percentage contributed by meat (Bailey & Peacock, 1988).

The oil palm is thought to have originated in West Africa, and fossil pollen grains of Miocene age have been found in the Niger Delta (Zeven, 1972). The species still grows wild, as well as in all stages of domestication, typically within 7° of the Equator. It is a light-loving species, and so is more characteristic of disturbed than of primary forest, especially in well-watered lowlands. The species has spread from the Congo into East Africa, especially in the wetter areas along the Great Rift, but the Arab slave trade was responsible for its wider spread to parts of the Indian Ocean coast.

Six key sites

Site 1 At Assirik, in the Parc National du Niokolo-Koba (PNNK) of Senegal, there is a small population of chimpanzees in a marginal environment at the edge of the species' range (McGrew *et al.*, 1981). No trace of oil palms was found over 4 years in the core study-area of 50 square kilometres, either in systematic transecting or in repeated searching. Further, analysis of over 800 specimens of chimpanzees' faeces over that period yielded no traces of oil palm (McGrew *et al.*, 1988).

However, contact between ape and oil palm cannot be ruled out altogether. During a survey of vegetation of the PNNK, Schneider & Sambou (1982) found oil palms growing in gallery forests elsewhere in the park. At least three of the sites (Badi, Niokolo-Koba, Simenti) that they visited also had chimpanzees recorded there. All three places are accessible to chimpanzees from Assirik by riverine routes. Thus the wide-ranging chimpanzees (Baldwin *et al.*, 1982) of Assirik may have eaten oil palms during their travels, but this remains to be shown.

Finally, no other species of nuts as consumed at more forested sites elsewhere were found at this hot, dry and open savanna site, but other kinds of palms such as *Borassus* were likely eaten by the chimpanzees (McGrew *et al.*, 1988).

Site 2 At Gombe on the eastern shore of Lake Tanganyika in Tanzania, is the best-known population of wild chimpanzees in the world (Goodall, 1968, 1986). Oil palms are abundant: Clutton-Brock & Gillett (1979) found them to be the sixth most common species of tree in the forest. On a line transect through several vegetation types, oil palms ranked joint fifth, making up 7% of the trees encountered.

The high-energy mesocarp eaten by the chimpanzees is the single most frequent type of food in their diet. It often stains their faeces a yellowish-orange. For 10 of the 12 months of the year, most specimens of faeces contained oil palm nuts or fibre, implying daily feeding on the species (Goodall, 1968, p. 184). The apes also eat the flower, pith, resin and cambium of the oil palm (Wrangham, 1975).

There is nothing unsavoury about the discarded nuts, which are eaten by sympatric olive baboons and bush pigs (Wrangham, 1975). Both forms crack the nuts open with their heavy molar teeth. Chimpanzees and baboons directly compete for oil palm nuts during the day, while the pigs scrabble for them on the ground at night.

Site 3 At Kasoje, in the Mahale Mountains of Tanzania, the evidence that chimpanzees do *not* eat oil palms is unequivocal. It has never been seen, though the apes have been fully observable at close range for over two decades (Nishida, 1990). Similarly, thousands of faecal specimens have been analysed without a trace of oil palms being found. Together the two data-sets provide the strongest negative evidence ever assembled in field studies of chimpanzees' diet.

Yet the apes could eat oil palms, as many groves of the trees are available (Nishida *et al.*, 1983). It is not that Kasoje's chimpanzees are especially choosey or averse to domesticated forms: they include six kinds of cultigens among the 198 species eaten, which is the most diverse repertoire of any population of chimpanzees known. Nor are they conservative in diet, as they have continued to expand their menu to include new items such as mangos (Takasaki, 1983). Finally, the oil palms at Kasoje are of proven palatability: from the same trees harvested by the other animals, the local Tongwe people take fruits to make cooking oil.

Site 4 At Lopé, in central Gabon, a still-wary population of chimpanzees is being studied in a large, undisturbed block of equatorial forest (Tutin *et al.*, 1991). Oil palms are probably less abundant than at Gombe: they made up only 2 of 350 trees found on a 10-metre-wide transect over 1 kilometre long (Williamson, 1988). This would mean that oil palms occur at a density of about 100 per square kilometre. Both oil palm fibre and undamaged nuts turn up in chimpanzee's faeces year-round (C. Tutin & M. Fernandez, unpublished data). Several genera (e.g. *Detarium, Panda, Saccoglottis*) of trees bearing nuts that are cracked open with hammers by chimpanzees elsewhere are present at Lopé, but the chimpanzees ignore them (McGrew *et al.*, 1992).

Site 5 At Tai, in southwestern Ivory Coast, is the largest remaining tract of rain-forest in West Africa. Oil palms are localised along three of the eight rivulets in the range of the chimpanzees being studied. However, the production of nuts is low, perhaps because of damage to the trees from the chimpanzees eating the other parts. At any rate, the nuts are virtually ignored by the apes, as only one faecal specimen in 25 months of study showed a sign of one having been eaten (C. Boesch, personal communication).

However, Tai's chimpanzees present the apex of non-human lithic technology in exploiting five other species of nuts (Boesch & Boesch, 1981, 1983, 1984*a*, *b*; Kortlandt, 1986). They use hammers of stone or wood and anvils of stone or root to crack open *Coula edulis, Detarium senegalense, Panda oleosa, Parinari excelsa* and *Saccoglottis gabonensis*. Despite the fact that all of the five species are probably harder shelled (Kortlandt, 1986) and that the hammering techniques for all species are equally efficient (Boesch & Boesch, 1983) the apes do not apply their skills to the vulnerable palm nut.

Site 6 At Bossou, in extreme south-eastern Guinea near the Nimba Mountains, is a small population of wild but tame chimpanzees (Sugiyama, 1984). They raid the crops of local villagers, and use both hammers and anvils of stone to crack open palm nuts to extract the kernels (Sugiyama, 1981; Sugiyama & Koman, 1979). The tools and work-sites of the human and non-human primates are indistinguishable; only their locations on or away from human paths allow them to be discriminated (Kortlandt, 1986; Kortlandt & Holzhaus, 1987).

It seems that the use of such elementary technology to exploit palm nuts is increasing at Bossou, as the apes find it harder and harder to subsist in a shrinking natural habitat (Kortlandt & Holzhaus, 1987). The mean number of hammers found at work-sites has tripled over 7 years of study, and all suitable palm trees (in terms of available stones, safe locations, gentle slope, etc.) are used. Despite this apparent pressure of maximal exploitation, neither Sugiyama nor Kortlandt found any evidence of other species of hard-shelled foods being cracked open, though the appropriate species of trees were likely to be available. Thus the technique appears to be highly selective.

Other sites

Surveying the full range of relations between chimpanzee and oil palm can be eased by making a 2 × 2 matrix of sites (see Table 1.2). Those places where oil palms are present versus absent can be opposed to those where palm nuts are eaten versus not eaten. This gives four cells, one of which is illogical and so can be ignored: absent oil palms cannot be eaten. In filling in the other three cells with data it is essential to include *all* known cases of both positive and negative evidence, to avoid inadvertent bias. This is easy enough for the 11 sites at which chimpanzees have been studied in the long term. It is more difficult for short-term studies, some of which are little more than anecdotes. (See Chapter 2 for meaning of long- versus short-term.) For example, Beatty's (1951) half-page note gives few background details to a single incident of tool-use. What follows here is meant to be exhaustive in coverage but not exhausting in length.

The simplest of the three possibilities is sites where oil palms are absent, and so chimpanzees cannot eat them. In addition to Assirik, there are four other long-term sites where this applies, in Uganda and Tanzania.

In the Budongo forest in western Uganda, Eggeling's (1947) extensive description of the vegetation did not include oil palms. The two main field studies done there, by Reynolds & Reynolds (1965) in 1962 and by Sugiyama (1968) in the mid-1960s, also made no mention of the species. However, this negative evidence is not likely to be conclusive, for several reasons. In each case the study lasted less than a year and was of subjects not tolerant of being continuously watched by human beings at close range. Apparently, neither study included analyses of faeces. The lists of species of plant-foods given are probably incomplete, as the larger of the two (the Reynolds's) numbered only 35 species. Thus, one cannot rule out oil palms in the diet of Budongo's chimpanzees, but the species is likely to be no more than a minor food, if present and eaten at all.

The Kibale forest, also in western Uganda, is better studied, being one of a handful of major study-sites in the world for the behavioural ecology of primates. The most extensive raw data on the site's vegeta-

Table 1.2. *Sites where oil palms are present or absent and chimpanzees do or do not eat them*

	Eaten	Not eaten
Present	Bassa (Liberia) BOSSOU (Guinea)[a] Cape Palmas (Liberia/Ivory Coast) GOMBE (Tanzania) Guinea (de Bournonville's study) Kindia (Guinea) Beatty (Liberia) LOPÉ (Gabon)	Belinga (Gabon) Beni (Zaïre) KASOJE (Tanzania) Kilimi (Sierra Leone) OKOROBIKÓ (Equatorial Guinea) Sapo (Liberia) TAI (Ivory Coast)
Absent		ASSIRIK (Senegal) BUDONGO (Uganda) Filabanga (Tanzania) Guinea (de Bournonville's study) KABOGO (Tanzania) KASAKATI (Tanzania) KIBALE (Uganda) Tiwai (Sierra Leone)

[a] Upper case indicates long-term studies.

tion are probably those given by Struhsaker (1975). Strip and quadrat counts of the trees yielded no specimens of oil palm (T. Struhsaker, personal communication). Study of the chimpanzees of Kibale was done by Ghiglieri (1984), who did not mention oil palms. However, his study shares some of the drawbacks of those at Budongo: the apes were only minimally habituated, no faecal analyses were done, and observation was largely focussed on fruiting fig trees. Only 20 species of plants were recorded as being eaten. Ghiglieri's results cannot rule out oil-palm-eating, but the evidence for the species' absence is strong.

In the Kabogo Point area in Tanzania, on the shore of Lake Tanganyika between Gombe and Kasoje, Azuma & Toyoshima (1961–2) studied wild chimpanzees for over a year. They recorded 38 kinds of plants being eaten by the apes, but these did not include oil palms, which were not present (S. Azuma, personal communication).

Similarly, both short-term (Izawa & Itani, 1966) and long-term (Suzuki, 1969) studies of the ecology of chimpanzees in the Kasakati Basin in Tanzania found no evidence of oil palms being eaten. A. Suzuki (personal communication) collected much faecal data, which were negative, and never saw oil palms in the forest away from settlements.

Kabogo and Kasakati are less than 30 kilometres apart. In his regional survey that included both places, Kano (1972, p. 64) reported that oil palms cultivated by local villagers were commonly found both on the lakeshore and inland. He found chimpanzees raiding palm nuts from a village less than 30 kilometres from Kasakati. Thus, chimpanzees at both sites may have met oil palms peripherally but not in the forest (T. Nishida, personal communication).

Of the short-term studies, de Bournonville (1967, table VIII) did a country-wide survey in Guinea lasting 4 months. He found that though chimpanzees' eating of oil palms was widespread, there were two adult male chimpanzees in captivity who refused palm nuts while accepting other fruits from their captors. He speculated that they might have come from areas where oil palms were absent and so were showing dietary conservatism.

The next simplest case is that in which the oil palm is there to be eaten but is ignored by chimpanzees. In addition to studies at Kasoje and Tai, there was a long-term study of chimpanzee diet done at Mt. Okorobikó in Equatorial Guinea (then called Rio Muni) by Jones & Sabater Pí (1971; Sabater Pí, 1979). Data on both predator and prey are sparse. No published mention of oil palms was made, although other species of plants used by humans were present and eaten by the chimpanzees. However, J. Sabater Pí (personal communication) confirms that oil palms were present on the lower fringes of the mountains, and that chimpanzees came into contact with them, but that there was no sign of oil palms being eaten by the apes.

Other, short-term studies gave evidence of chimpanzees abstaining from available palm nuts. In the Kilimi area of northern Sierra Leone, oil palm fruits were eaten by white-collared mangabeys and by Guinea baboons but not apparently by sympatric chimpanzees (Harding, 1984). However, R. Harding (personal communication) adds that most oil palms grew near villages, which the shy chimpanzees were unwilling to approach.

In the Sapo (or Sarpo) National Park in eastern Liberia, chimpanzees were studied for 3 months in 1981 (Anderson *et al.*, 1983). The apes used pieces of laterite as hammers to smash open four kinds of nuts, but these did not include palm nuts. The species was present in the forest, but no signs of chimpanzees' eating it were found (E. Williamson, personal communication). However, this was little more than a pilot study, and longer, systematic research at Sapo is needed.

At Belinga, in north-eastern Gabon, both chimpanzees and gorillas were studied intermittently in 1980–2 (Tutin & Fernandez, 1985). Oil palms were present, but no evidence from direct observation, feeding traces or faecal specimens indicated their being eaten by the apes. However, the data-set for chimpanzees was small.

Finally, one attempt at field experimentation with palm nuts produced negative results: Kortlandt (1967) five times offered a bunch of oil palm fruits to wild chimpanzees in the Beni Chimpanzee Reserve in north-eastern Zaïre. The stimulus was left on a path along which the apes passed, and their responses were recorded from hiding. Occasionally, they showed mild curiosity but mostly they detoured around the fruits. The chimpanzees were not averse to all such gifts, however, as they accepted bananas and papayas.

There are at least six other places in Africa where chimpanzees are known to eat palm nuts, in addition to the long-term study-sites of Bossou, Gombe and Lopé. One is Cape Palmas, on the coast near the present-day border between Ivory Coast and Liberia. Savage & Wyman (1844), in arguably the first published account of the natural history of the chimpanzee, recorded that oil palms were abundant and freely eaten by chimpanzees. They made no mention of tools being used to process the palm nuts, but one other species of hard-shelled fruit was cracked open with stones. Beatty's (1951) brief note supplied the first record of tool-use and oil palms but also said nothing about other species. The location of Beatty's chance encounter was not given; at the time he could not have known how useful the information would be decades later.

In Guinea, the first field-study of wild chimpanzees, done near Kindia in western Guinea, found that chimpanzees ate palm nuts (Nissen, 1931, p. 57). Later, as mentioned above, de Bournonville (1967, p. 1248), working mainly in the north of Guinea, found that chimpanzees commonly ate palm nuts.

The best *behavioural* data on chimpanzees using tools to crack open palm nuts come from a free-ranging group released onto an offshore island in Liberia (Hannah, 1989; Hannah & McGrew, 1987, 1990) (see Figure 1.1). Sixteen wild-born apes aged 5–20 years were set loose after varying lengths of time in captivity. When one adult female unexpectedly began to crack open palm nuts using a piece of concrete as a hammer and the cemented water dispenser as an anvil, 12 of the others followed suit. Eventually, the chimpanzees spontaneously carried both hammers and nuts to new sites in the forest and made use of tree branches, fallen logs and mangrove roots as anvils. Whether or not the

Figure 1.1. One adult female watches another use a hammer stone and anvil to crack open oil palm nuts, Bassa Islands, Liberia. (Courtesy of A. Hannah.)

habit spread through the group by observational learning is unclear; an alternative explanation is that the performance by the female prompted in the others long-dormant memories retained from their wild upbringing. *All* oil palm eating and nut cracking in wild populations was already established when observers began study, so no natural cases of innovation are known.

These are the findings on chimpanzees as predators on oil palm nuts. I will now tackle the question posed in this chapter's title, as taken from Benedict's (1935) book of the same name.

Non-human culture?

To what extent do species other than humans possess something akin to human culture? Are natural scientists over-interpreting their findings when they assert, for example, that dialects in bird-song meet all the criteria of fully-fledged culture (Mundinger, 1980)? Put another way, would socio-cultural anthropologists allow themselves to be persuaded by *any* data which the animal behaviourists might produce? Can ethology address ethnology? Or, as in the 'pongo-linguistic' controversy of the 1970s on the nature of language, is the phenomenon of culture thought by *definition* to be uniquely human?

One set of answers to these questions lies in the issues of *definition*, which bring many problems. For example, it is all very well to produce theoretically satisfying definitions, but if these cannot be translated into operational ones, then empirical testing is impossible. This chapter is not about definition, but Chapter 4 is. Here, it is assumed that satisfactory definitions exist or can be devised. Instead this chapter is about a method, and its aim is to try a new sort of exercise, at least as applied to apes. It involves focussing on a particular habit, that is, chimpanzees preying on oil palms, in order to compare its incidence in all suitable populations, that is, those for which data are available. The crucial next step is to assess on the basis of comparison of these data the usefulness of this method for understanding the origins of culture.

Before going on, it seems advisable to review briefly other approaches to chimpanzee culture which have been advanced. The earliest and still most common of these was for the observer to describe the ingenious and sometimes astonishing acts of a single population of apes. These patterns were seen as being the results of individual learning, even if their origins could not be specified. The first examples were seen in captive apes, either in groups (Köhler, 1927) or in individuals reared in human homes (Hayes & Hayes, 1954). This approach continues to yield fascinating descriptive results, especially as captive environments become richer and more stimulating (de Waal, 1982). Similar results emerged in the 1960s from the wild, especially from long-term studies such as those of the Kasakela community at Gombe studied by Goodall (1973).

The problem with all such studies is that with a sample size of one, be it an individual, group or population, there is no variance. One can always argue that a unique set of conditions has prompted an idiosyncratic response, whether ontogenetically or phylogenetically programmed. For example, if a group of chimpanzees in a zoo knows how to use fire, they may have been shaped that way by the impinging acts of their keepers and the public (Brink, 1957). Or, if only one population of chimpanzees preys upon a painfully stinging kind of ant, then this may reflect the unusual make-up of a marginal habitat were subsistence is harsh (McGrew, 1983). That is, extraordinary circumstances may prompt similarly extraordinary responses. There is no need in these cases to call upon any social factor, which is assumed to be a necessary, but not sufficient, condition for culture.

A more useful approach is that of limited comparison. This depends on having at least two comparable sets of data, and such raw material for chimpanzees became available from the 1970s onwards. (A special sort of two-way comparison, that of captive versus wild, began earlier, but such a gross contrast yields largely self-evident results.) The most common paired comparison has been that of two neighbouring populations: Gombe versus Kasoje (McGrew & Tutin, 1978; Nishida *et al.*, 1983). A few others have gone further afield (Hladik, 1977; Baldwin *et al.*, 1981). Surprisingly little has been done along these lines with captive apes. For example, several zoological gardens around the world have offered artificial termite mounds to groups of confined chimpanzees (Nash, 1982), but no comparative data on their use seem to have been sought.

There are several drawbacks to such limited comparisons. One also follows from the constraints of small sample size: an apparently notable contrast may merely reflect an anomaly in one of the two sets being compared. Another is that selective comparisons of animal prey may emphasise a few differences at the expense of many more similarities of plant-foods. Conversely, approximate and general rather than systematic and precise comparisons may obscure differences which lurk behind vague generalisations of similarity. Most problematic, however, is that comparisons of the subjects may really be comparisons of the scientists, especially if their methods of data collection differ. For example, if in comparing diets one set of data is based on faecal sampling (which is biassed against food-items of only soft tissue that are fully digested) and the other is based on direct observation (which is biassed against inconspicuous events, except in totally observable subjects), it would not be surprising to find apparent but false differences. This is not an idle complaint; it applies to most possible pairings of data from long-term field studies of chimpanzees.

Put another way, the ideal two-way comparison involves a close match-up, with all or as many as possible of the independent variables being controlled for. Contrasts found in two sets of subjects will be more convincing if done by the same investigators using the same methods, etc. This is rarely achieved. Another approach is to randomise out these variables by using many sets of data, on the assumption that differences in technique, or definition, or sampling, or precision, or any other noise in the system, will pale to insignificance. Any phenomenon which 'survives' this treatment is likely to be a robust one. For example, if all wild chimpanzees build arboreal sleeping-platforms by inter-weaving the terminal branches of trees, regardless of when and where they were studied, for how long, by whom, using whatever system of recording, then it seems safe to conclude that one is dealing with a universal aspect of chimpanzee nature. The rub is that many sets of data are needed for this sort of exercise, and this really became possible only in the 1980s, as fieldwork, especially in West Africa, advanced. Of course, in this approach one is obliged to use all available sets of data that meet the criteria of suitability, so that no bias through selectivity creeps in.

So, what do these comparative data on chimpanzees and oil palms tell us about culture and non-human nature? The first sort, of absent oil palms not being eaten, say little except that chimpanzees can survive readily without this species of prey. This is hardly surprising for such an omnivorous forager. Methodologically, it emerges that it is not always easy to say something so simple as whether or not even a highly recognisable species like the oil palm is available. Presumably this reflects an understandably greater concern by field-workers with what is present and used, not absent or ignored.

The second sort of evidence, of accessible oil palms being ignored by chimpanzees, is more interesting. Although the principle of 'absence of evidence is no evidence of absence' may apply to Okorobikó, this caveat hardly explains away Kasoje. The population of chimpanzees at Kasoje is large, healthy, enterprising and well-known. The simplest hypothesis to explain why they ignore such a useful food-stuff is that it is not within their body of social tradition. In other words, unlike other items in their diet the use of which is passed on by social learning from one generation to another, the oil palm is apparently not seen as being edible. The apes' ignorance seems to be a cultural accident, not something dictated or even influenced by the natural environment.

One cannot entirely rule out alternative hypotheses, of course, however far-fetched. It could be that some as yet unrecognised competitor for Kasoje's oil palm nuts makes their use by those chimpanzees uneconomical. Or it could be that Kasoje's chimpanzees have to hand some other food-item which is preferable to oil palm nuts on all counts. However, these alternative explanations seem forced and unlikely, given the range of data in Tables 1.1 and 1.2. More likely is that Kasoje's oil palms were close to villages where chimpanzees feared to venture until relatively recently.

It is the third type of evidence, that of differences between populations of chimpanzees which make use of oil palms, that is most impressive. The apes at Gombe and Lopé eat the outer pulp of the fruit but discard the inner nut with its kernel. Bossou's chimpanzees go beyond this to eat the embedded kernel by using technology to gain access to it.

At both Lopé and Gombe there are ample stones available for use as hammers and anvils (D. Collins, unpublished data; McGrew *et al.*, 1992), but these are ignored. Even having the technical skills, as is the case at Tai, does not mean that they will be applied in all cases. What seems most likely to account for the data from Bossou is that a cultural innovation in the form of hammer and anvil has become part of a traditional repertoire. This conclusion is further strengthened by the fact that all known cases of hammer-stones being used to crack open hard-shelled nuts come from a relatively small part of the range of the chimpanzee (Struhsaker & Hunkeler, 1971; Teleki, 1974). The habit seems to be limited to Guinea, Ivory Coast, Liberia and Sierra Leone (Whitesides, 1985). This is *prima facie* evidence for limited cultural diffusion, and further detailed study of local variations in customs needs to be done (Kortlandt & Holzhaus, 1987). Only then will the richness of continuity and discontinuity in technique and performance come to light.

Again, sceptics may be able to offer non-cultural explanations to account for these comparative data. For example, it may be that oil palm nuts vary in hardness across the range of the species, such that some kinds lend themselves more easily to being smashed open. However, grasping at such straws has an air of desperation about it (cf. Harris, 1985). If the same findings cited above came from a range of *human* societies across Africa, we would not hesitate to call the differences cultural.

2

Studying chimpanzees

Introduction

The chimpanzee has been studied more intensively and extensively than any other species of non-human primate in Africa.

Little or nothing is known scientifically about the chimpanzee in most of the countries of Africa in which the species occurs.

These two statements seem paradoxical, but both are likely to be true. As regards the first, intensive studies of up to 30 years' duration continue at several sites, and matrilineal kinship in some communities is known for four generations (Goodall, 1986). Life-histories of individuals whose longevity approaches that of pre-industrial human beings are accumulating: a male (Goblin) born in 1964 achieved alpha-rank 20 years later, but may still have as many years of life left (Goodall, 1986). Literally thousands of hours of close-up observation have been recorded on some *individuals* (Goodall, 1986). Extensively, chimpanzees have been studied throughout the geographical range of the species. This stretches from the Mahale Mountains of Tanzania in the south-east to Mont Assirik in Senegal in the north-west (Lee *et al.*, 1988). The straight-line distance between the two points is a staggering 5300 kilometres or 3300 miles. (For comparison's sake, the distance from New York to Los Angeles is only about 4000 kilometres.)

Considering the second statement, the chimpanzee could occur, at least in principle, in 29 countries in Africa (Lee *et al.*, 1988; McGrew, 1989*b*) (see Table 2.1). That is, at least part of each of these nations contains suitable habitat in terms of vegetation and rainfall and borders on at least one other country *known* to have wild chimpanzees. (Of course, political boundaries often have little to do with biotic ones, but

Table 2.1. *African countries which* could *have wild chimpanzees (Lee* et al., *1988)*

Far western (*P. t. verus*)	Central-western (*P. paniscus, P. t. troglodytes*)	Eastern (*P. t. schweinfurthii*)
IVORY COAST[a]	EQUATORIAL GUINEA	TANZANIA
SENEGAL	GABON	UGANDA
GUINEA	ZÄIRE	Sudan
SIERRA LEONE	Cameroon	Zaïre
Liberia	Central African Republic	Burundi
Mali	Congo	Rwanda
Ghana	Nigeria	(Zambia)[b]
Guinea-Bissau	Angola-Cabinda	(Malawi)
(The Gambia)		(Kenya)
(Burkina Faso)		
(Togo)		
(Benin)		
(Niger)		

[a] Upper case indicates long-term studies.
[b] Brackets indicate chimpanzees absent or current presence unconfirmed.

the former usually determine opportunities for field-researchers more than do the latter.) Of these 29, in only eight have there been long-term studies (see below for details), and in only a further *eight* has there been any kind of study of chimpanzees, even the most minimal survey (McGrew, 1989*b*).

The aim of this chapter is to outline the study of chimpanzees, both actual and potential, in terms of when, where, who and how. Field-study will be emphasised but, when pertinent, material on captive apes will be added.

Development of chimpanzee research

Studies in nature

Several detailed accounts of the discovery and early contacts with chimpanzees are already in print (Yerkes & Yerkes, 1929; Morris & Morris, 1966; Reynolds, 1967; Hill, 1969). The reader is referred to these, and what follows is only a sketch.

Apart from early, misconceived attempts (Garner, 1896), the first scientific field-study was that of Nissen (1931) in Guinea. It remains an impressive effort. In only 4 months he noted data on daily activity cycles, diet, vocalisations, and even social life. He found no tool-use, but his first-hand findings on nests and drumming were the first on habitual use of objects by wild apes. By modern standards his report was wordy and non-quantitative, but his field-notes, many published verbatim, remain models of diary-style description.

After a gap of 30 years there was an explosion of field-studies of chimpanzees in the early 1960s. In 1960, Kortlandt (1962) went to what is now Zaïre and began observations of chimpanzees visiting a plantation. This was the start of a series of short studies which opted for breadth across the distribution of the species. For example, he was the first to study apes in both eastern and western Africa. Kortlandt was also the first to devise experimental tests of tool-use with wild subjects (Kortlandt & Kooij, 1963).

Also in 1960, Goodall (1968, 1986) began the longest-running field-study of any species of ape in nature, in what is now Tanzania. Her study at Gombe in woodland on the eastern shore of Lake Tanganyika has compiled more hours of close-range observation than all other studies put together, and its popular impact has been immense (Goodall 1967, 1971, 1990). For most people the chimpanzees of Gombe equal The Wild Chimpanzee. Notably, as Goodall has stressed again and again, new findings continued to emerge in the third decade of research. More to the point of this chapter, Goodall (1964) was the first to find tool-use and tool-making in a natural population of non-human primates.

In 1961, Imanishi's Kyoto University Anthropoid Expedition also began field research on chimpanzees in woodland and forest in western Tanzania (Azuma & Toyoshima, 1961–2). Study began on the eastern shore of Lake Tanganyika at Kabogo Point, then moved inland to Kasakati, Filibanga and Ugalla. Since 1965, however, virtually continuous research has been under way further south in the Mahale Mountains, as coordinated by Nishida (1979, 1990) of the University of Tokyo (now Kyoto). Many types of tool-use have been seen at Mahale (Nishida & Hiraiwa, 1982).

In 1962, studies began in Uganda when Reynolds & Reynolds (1965) undertook the first study of chimpanzees in evergreen forest, at Budongo. Sugiyama (1968) followed this with a further 6 months of study in 1966–7, as did Suzuki (1971) for 17 months in 1967–8. Later data on the chimpanzees of Budongo have been intermittent (Albrecht, 1976), and study of chimpanzees in Uganda has now shifted west to Kibale (Ghiglieri, 1984, 1988; Isabirye-Basuta, 1988; Conklin &

Figure 2.1. Adult male eats figs, Assirik, Senegal.

Wrangham, 1991). None of the early studies at Ugandan sites yielded evidence of habitual tool-use by chimpanzees.

Finally, in 1963, Sabater Pí (1979) began the first field-study of central-western chimpanzees in what is now Equatorial Guinea. Several forested sites were worked, but he focussed efforts on the Okorobikó Mountains, where western gorillas also occurred. The two species of ape were studied together for the first time (Jones & Sabater Pí, 1971). More pertinent here, a new kind of tool-use was seen, in which sticks were used to dig up termites for food (Jones & Sabater Pí, 1969).

After another quiet period, four long studies all began in 1976. In February, a team from the University of Stirling began a 4-year study at Mt. Assirik, in the Parc National du Niokolo-Koba in Senegal (McGrew *et al.*, 1981) (see Figure 2.1). There, a group of chimpanzees at the far north-western extent of the species' range lives in a savanna ecosystem, and uses a variety of tools made from vegetation (McGrew *et al.*, 1979*a*).

Following earlier surveys (Struhsaker & Hunkeler, 1971), the Boesches (Boesch, 1978) began in September an ongoing study of the chimpanzees of the Tai forest, Ivory Coast. At first they concentrated on hammers of wood and stone used by the chimpanzees to crack open

Table 2.2. *Long-term studies of wild chimpanzees*

Site	Country	Dates	Provisioned[a]	Key source
Assirik	Senegal	1976–9	No	McGrew *et al.* (1988)
Bossou	Guinea	1976–	Yes	Sugiyama (1989)
Budongo	Uganda	1962–8	No	Suzuki (1971)
Gombe	Tanzania	1960–	Yes	Goodall (1986)
Kabogo Point	Tanzania	1961–3	No	Azuma & Toyoshima (1961–2)
Kasakaki	Tanzania	1964–5	No	Suzuki (1969)
Kasoje[b]	Tanzania	1965–	Yes	Nishida (1990)
Kibale	Uganda	1976–	No	Ghiglieri (1984)
Lopé	Gabon	1983–	No	Tutin *et al.* (1991)
Okorobikó	Equatorial Guinea	1963–9	No	Sabater Pí (1979)
Tai	Ivory Coast	1976–	No	Boesch & Boesch (1990)
Tenkere	Sierra Leone	1990–4	No	Alp (1993)

[a] 'Provisioned' refers to the practice of giving artificial food to wild subjects.
[b] Kasoje is used collectively for several studies in the Mahale Mountains.

nuts, but since then they have extended their study to general behavioural ecology (Boesch & Boesch, 1989, 1990).

In November, Sugiyama (1989) began an intermittent study of the chimpanzees of Bossou, near Mt. Nimba, Guinea. The site was first used 10 years earlier by Kortlandt and his colleagues (Dunnett *et al.*, 1970; Albrecht & Dunnett, 1971). The community of apes is small and isolated but has an extensive repertoire of tool-use and diet (Sugiyama & Koman, 1979, 1987). In December, Ghiglieri (1984, 1988) began a 22-month study in the Kibale forest. He worked mostly at Ngogo, in 1976–8, and again in 1981. His study of chimpanzees was socio-ecological, and capitalised on vigils at popular fruiting trees.

The most promising candidates for productive, ongoing field studies of wild chimpanzees seem to be in eastern Zaïre along the Ishasha River, begun in 1989 (Steklis, 1990; Sept, 1992), and in northern Sierra Leone, in the Kilimi-Outamba area, begun in 1990 (R. Alp, unpublished data).

Only 12 studies have been long-term, that is, have lasted 12 months or longer, but this is still more than for any other species of non-human primate (see Table 2.2).

Studies in captivity

Scientific study of the behaviour of captive chimpanzees began before field-studies, and the first two set high descriptive standards. Kohts (1935) noted the psychological development of a young male called Joni between 1913 and 1916. Twelve years later she kept similar records of her son Roody. The resulting comparative monograph was published in Russian, but happily there is a 55-page summary in English. The ape showed a rich range of tool-use, including drawing, but no numerical analyses were done on the data.

Köhler (1927) made the first study of socially-living chimpanzees, including adults, in 1913–17. He collected data from nine apes housed at the Anthropoid Station of the Prussian Academy of Science on Tenerife, Canary Islands. The results derived mostly from a series of ingenious experiments testing chimpanzee intelligence. Most of these involved tool-use, and the paradigms are still in use almost 80 years later. Though Köhler's conclusions have sometimes been re-interpreted (Chance, 1960), his insights into the chimpanzee mind remain unrivalled (Beck, 1977).

Kohts's study has been replicated several times, with variations on the following theme: a human couple, at least one of whom is a psychologist, rears an infant chimpanzee in their home, often in the company of their children. The aim is to probe the limits of ape adaptability by holding constant all environmental variables, thus allowing a controlled comparison of human and non-human nature (Kellogg, 1969).

There are several well-known examples. Kellogg & Kellogg (1933) raised a female chimpanzee, Gua, with their son, Donald, for 9 months. The human infant was only 2.5 months older than the chimpanzee, so the two were exhaustively compared on a battery of 28 psychological tests and experiments. The Hayes's study of a young female chimpanzee, Viki, became best known for their limited success in teaching her to say four words (Hayes, 1951). Their published papers provide the most useful body of findings of any of the home-rearing studies (Hayes & Hayes, 1952, 1954). Sadly, Viki lived for only 6 years.

The longest-running study was that of another female chimpanzee, Lucy (Temerlin, 1975). For the first 13 years of her life she lived in a human home, but in 1977 she joined a colony of chimpanzees being rehabilitated into the wild in The Gambia (Carter, 1981, 1988). She died there in 1987. Lucy was taught many signs in American Sign Language, but more startling was her spontaneous tool-use: she was inclined to fix herself a martini, leaf through *National Geographic*, and then masturbate with a vacuum cleaner!

Later studies of captive groups of chimpanzees are notable for their scarcity. Two have made a great impact, however: Menzel (1974) released a group of eight wild-born youngsters into a 3.6-hectare enclosure at the Delta Regional Primate Research Center in Covington, Louisiana. They matured there over the course of 6 years in naturalistic surroundings. The planned research was elegantly experimental, but again and again the apes showed surprising use of instruments, such as inventing and elaborating upon the use of ladders (Menzel, 1972, 1973; McGrew *et al.*, 1975) (see Figure 2.2).

In 1971, the chimpanzee colony at Burgers' Zoo in Arnhem, Netherlands, was created (van Hooff, 1973). By 1974 the population was established, and up to 30 individuals have lived ever since in a 1-hectare enclosure and adjoining building. The superior conditions have allowed a stream of studies of rich social life (de Waal, 1978; Adang, 1986) and a superb popular book (de Waal, 1982). Unfortunately, no detailed account of tool-use has yet appeared, though anecdotal snippets are fascinating.

Apart from the studies of pongo-linguistics (see below), most other studies of confined chimpanzees make sorry reading. Whether in zoos or laboratories, the subjects typically live alone or in small groups in cramped quarters with limited furnishings and diet. Many, especially those who were hand-reared, are behaviourally disordered or lack social skills. They may need prolonged therapy if they are to live sane lives (Fritz & Fritz, 1979). Any performance of tool-use in experimental testing is therefore remarkable, and is probably positively correlated with the socio-ecological validity of the captive environment. Compare, for example, the long training needed to elicit even the simplest stick-use in deprivation-reared chimpanzees (Birch, 1945) with the spontaneous making and using of stick-tools by chimpanzees in an enlightened zoo (Nash, 1982).

More stimulating lives have been led by most of the chimpanzees in studies of acquired language. This is not the place to review this controversial research, but much of it entails object manipulation. Premack's (1971) artificial language used plastic pieces put onto a magnetised slate, and later non-linguistic items ranged from cut-up photographs (Premack, 1975) to standardised test papers (Premack *et al.*, 1978). The most famous sign-language-using chimpanzee, Washoe, was raised in a caravan, and many household objects were used as test-

Figure 2.2. Tool-use in escape from an outdoor enclosure, Delta Primate Center, USA. Left foot of upper individual is resting on a piton jammed into the fence. (Courtesy of P. Midgett.)

items from the start (Gardner & Gardner, 1969). Sometimes objects were crucial for rigorously testing specific abilities such as transfer of ideas across the senses (Fouts *et al.*, 1974). Computer-automated study of chimpanzees using an artificial language, Yerkish, first required only a keyboard (Rumbaugh, 1977), but later, richer studies have included real-world objects in studies of symbol-use and communication (Savage-Rumbaugh *et al.*, 1978; Savage-Rumbaugh, 1986). Overall, however, pongo-linguists have used tools only as means to help gain access to the chimpanzee mind.

Sites of study

What follows is an overview of the geographical spread of research on chimpanzees, again stressing fieldwork. At first glance, the summed area of the 29 countries with known or likely wild populations is immense. Even the distribution of study-sites exceeds that of any other species of non-human primate.

Eastern chimpanzees

Pan troglodytes schweinfurthii is the best-studied of the three geographical races, though long studies have been done in only two of the nine potentially suitable countries.

In Tanzania, chimpanzees occur only in the far west along the eastern shore of Lake Tanganyika and in its hinterland. Research has been done from Gombe (almost at the border with Burundi) in the north, to Kasoje in the Mahale Mountains peninsula in the south. A thorough regional survey covering about 20 000 square kilometres was done by Kano (1972), but this is now 20 years old and needs updating. Of particular interest for further study is the Ugalla area inland to the east of the Lake. Surveys (Itani, 1979; Moore, 1986; Nishida, 1989) suggest that it may be the most arid environment in which wild chimpanzees survive, and if logistical obstacles can be overcome, more research should be done.

In Uganda, research began at Budongo but has shifted to Kibale, as outlined above. Further studies are under way at Kibale (Isabirye-Basuta, 1988, 1989; Conklin & Wrangham, 1991; Wrangham *et al.*, 1992) and new studies are beginning at Bwindi-Kayonza (Butynski, 1986). The extent to which chimpanzees survive to be studied in Uganda's other forest-blocks, where earlier workers found them (Haddow, 1958; Stott & Selsor, 1959; Reynolds & Reynolds, 1965), remains to be seen, but Reynolds has now returned after a gap of almost 30 years to revive the studies.

In Zaïre, chimpanzees were studied early in the Western Rift Valley by Kortlandt (1962). He worked for three periods totalling 10 months in 1960–4 at Beni, in Kivu province. There he took advantage of the crop-raiding habits of the local chimpanzees to set up observation hides in a plantation. He put out many objects, the most dramatic of which was a stuffed leopard with a chimpanzee doll in its paws. This prompted vigorous use of weapons by the apes (Kortlandt, 1967). In the driest habitat yet recorded for wild chimpanzees, along the Ishasha River in Zaïre, Steklis (1990) and his co-workers have begun periodic ecological study of a small population. This has already yielded valuable data for palaeoanthropological modelling (Sept, 1992). In principle at least, it is important to do comparative studies at sites such as Kahuzi-Biega where chimpanzees and gorillas co-exist (Yamagiwa *et al.*, 1988).

In Sudan, chimpanzees have been recorded in the extreme south of the country, in the Equatoria region just north of the border with Congo (Kock, 1967). Updated information is needed.

In Burundi, some chimpanzees remained in the Teza forest until recently (Verschuren, 1978), but their current status is unknown. New studies are under way (J. Goodall, personal communication).

In Rwanda, there are still chimpanzees, at least in the Nyungwe Forest Reserve (E. Williamson, personal communication). A survey is needed.

In Kenya, chimpanzees could in principle live in the fragmented forests along the west side of the Eastern Rift Valley, such as Kakamega, but they do not.

However, it is the southern boundary of the species' range that poses the riddle. There is no obvious present-day barrier that explains why chimpanzees fail to extend down the Western Rift Valley into Malawi and Zambia; however, the periodically inundated Rukwa gap (R. Grove, personal communication), may have inhibited them in the past. Apart from a controversial case in Malawi (Mitchell & Holliday, 1960; Hill, 1963; Benson, 1968), no chimpanzee has been seen in either country, though there are habitats in northern Zambia that are suitable (Ron & McGrew, 1988).

Central-western chimpanzees

Pan troglodytes troglodytes is the least studied of the three geographical races, though it lives in the part of the species' range which is most likely to be the ancestral homeland: equatorial forest.

In Equatorial Guinea (then Rio Muni), Jones & Sabater Pí (1971) studied unprovisioned but crop-raiding chimpanzees at three sites: Mt. Alen, Abuminzok-Aninzok and Mt. Okorobikó. Their study lasted 16 months in 1967–8, but Sabater Pí (1979) worked over a longer period (1963–9) at Okorobikó. From there he gave details of chimpanzees' use of sticks to get termites (Sabater Pí, 1974). No more recent information on the status of the chimpanzees has since appeared.

In Gabon, studies of apes lagged behind those of monkeys. Hladik's (1973, 1977) year-long study in 1971–2 of a group of chimpanzees released onto an island in the Ivindo River was the only one until the 1980s. Then, the most comprehensive nation-wide survey of any primate was done, lasting 27 months (Tutin & Fernandez, 1984). Intensive study was based at Belinga in the north-east, from whence data on diet (Tutin & Fernandez, 1985) and tool-use (McGrew & Rogers, 1983) emerged. Since 1983 a comparative socio-ecological study of chimpanzees and gorillas has been under way at the Station d'Etudes des Gorilles et Chimpanzés in the Lopé-Okanda Reserve (Tutin *et al.*, 1991).

In Cameroon, primates and primatologists abound, but chimpanzees were long ignored, though known to be present (Gartlan & Struhsaker, 1972; Mitani, 1990). Only a 2-month study in 1984–5 at Campo Reserve has been done (Sugiyama, 1985). It yielded variations on the theme of sticks used as tools to get termites, as described to the south in Equatorial Guinea (Sabater Pí, 1974) and Gabon (McGrew & Rogers, 1983).

In Central African Republic, in contrast, little research on any primates has been done. However, recent surveys in the far south-west, Haute Sanga prefecture, showed that chimpanzees and gorillas occur sympatrically there (Carroll, 1986, 1988; Fay, 1989; Fay & Carroll, 1990).

In Congo, no primatological research has been done until recently (Fay & Carroll, 1990), but chimpanzees have long been known to live there (Spinage, 1980). The best prospect for study may be the Odzala National Park and the Ndoki forest in the north.

In Cabinda, the enclave of Angola in Congo which is on the north side

Figure 2.3. Wild bonobos at Lomako, Zaïre (Courtesy of A. & N. Badrian.)

of the Zaïre River, no recent data are available on chimpanzees (Lee *et al.*, 1988).

In Nigeria, little is known of the current status of the chimpanzee, although some populations survive (Lee *et al.*, 1988). A survey of distribution and numbers is needed on the east and west sides of the Niger River, which may be the zoo-geographical barrier between the central-western and far western chimpanzees.

Studies of bonobos (or pygmy chimpanzees) were late in starting by comparison with those of chimpanzees. Nishida's (1972) survey of the Lac Tumba region was followed by a long study there by Horn (1980). These produced few observations but valuable ecological data on a heavily hunted population of apes. Kano (1979, 1984) got more promising results from a survey done with Nishida further south in 1973. The transition from survey to ongoing study came in 1974, with the emergence of two sites: Lomako and Wamba (see Figure 2.3). It was Badrian & Badrian (1977) who made the break-through in the north with their 11-month study in 1974–5 in Equateur Region. This led to the setting up of a permanent research station in the Lomako forest (Susman, 1984), where three unprovisioned (see below) groups have since become increasingly observable (White, 1992). To the south-west, a parallel study of five, heavily provisioned groups is under way at Wamba (Kano & Mulavwa, 1984; Kuroda, 1980). A fourth place, Yalosidi, even further south, was studied ecologically first by Kano (1983) and then by Uehara (1990). Despite much attention being paid at all sites, tool-use so far seen in wild bonobos is limited to the use of leafy twigs as rain-shields (Kano, 1982).

Western chimpanzees

Pan troglodytes verus was the first of the three geographical races to be studied (by Nissen), and it has also been the most heavily exploited for export. Yet relatively little is known of its natural behaviour.

In Ivory Coast, the only study of chimpanzees has been that of the Boesches at Tai (see above). Tai is the only substantial block of forest left in the country, but surveys elsewhere show chimpanzees to exist in Azagny and Comoé National Parks. Comoé, in the far north, is of great potential interest, as chimpanzees seem to be on the west side of the river but not the east (Geerling & Bokdam, 1973). Thus a natural barrier presents the potential for ecological study with a 'control' condition lacking apes.

In Senegal, a few chimpanzees are present in Oriental province, in the south-east, and the only studies have been by McGrew *et al.* (1981) and a follow-up by Bermejo *et al.* (1989). Further surveying to the south and east of the Parc National du Niokolo-Koba is needed.

In Guinea, Kortlandt (1986) did surveys and short studies spanning 1960–86 (seven trips totalling 7 months in the field). These short studies concentrated on two sites: Kanka Sili and Bossou. Sugiyama's (see above) intermittent study has been at Bossou, while Albrecht & Dunnett (1971) concentrated their 6-month study at Kanka Sili. They shot 12 000 metres of cine film from hides, much of it showing weapon-use in response to artificial stimuli such as a stuffed leopard. After Nissen's early survey (see above), the next was by de Bournonville (1967) further north in the Fouta Djallon. In 4 months he travelled throughout an area of 90 000 square kilometres, including a foray into Niokolo-Koba in Senegal. In principle, at least, Guinea presents the opportunity for rigorous study of chimpanzees in a cline from wet forests in the south to dry woodlands in the north.

In Sierra Leone, Harding (1984) did a 6-month survey of wildlife in the Kilimi-Outamba region, in the north along the border with Guinea.

A small population of chimpanzees occurs there and a follow-up study is under way at Tenkere in Outamba (Alp, 1993). In the south-west, Whitesides (1985) reported chimpanzees on Tiwai Island using stone tools to open *Detarium* nuts.

In Liberia, only one study of wild chimpanzees has been done. Anderson *et al.* (1983) spent only 2 months in the Sapo National Park, but found both meat-eating and stone tools being used to crack open four species of nuts.

In Mali, a survey of large mammals in 1972–4 revealed chimpanzees to be in the far west of the country (Sayer, 1977). Moore's (1985, 1986) survey showed that chimpanzees still exist there, along the Bafing River, in a savanna habitat. This too needs further study.

In Ghana, no research on chimpanzees has been done, but they were once common (Lee *et al.*, 1988), and a few chimpanzees may survive in Bia National Park (Jeffrey, 1975).

In The Gambia, there are no wild chimpanzees left, but released chimpanzees range freely on islands in the Gambia River (Carter, 1981, 1988).

In Benin and Togo, the paired countries of the Dahomey Gap, chimpanzees once occurred (Burton [1966], p. 329; Cornevin, 1969, p. 19), but they are feared now to be extinct (Lee *et al.*, 1988).

In principle, chimpanzees may survive in riverine forests in Burkina Faso and Niger, but no surveys have been done. The same uncertainty applies to Guinea-Bissau, which still has some forests, and so seems more promising (Lee *et al.*, 1988).

Captive chimpanzees

Few studies of confined chimpanzees have been done in their home continent of Africa. Research in zoological gardens is limited to the anecdotal (Brink, 1957), and the few primatological laboratories found in countries with wild apes are largely biomedical, not behavioural. Rehabilitation projects have been concerned mostly with welfare (Brewer, 1978) and not research, though this is changing (Ron & McGrew, 1988; Hannah & McGrew, 1991).

Instead, research on captive chimpanzees is concentrated in North America, Japan and western Europe. In the USA, two of the seven Regional Primate Research Centers have specialised in apes. Menzel's work (1974) was done at Delta in Louisiana, and many scientists have worked at Yerkes, in Atlanta. Of the other large colonies in Arizona,

Georgia, New Mexico, New York and Texas, only the first and last support behavioural research (Fritz & Fritz, 1979; Maki *et al.*, 1989). The Holloman Air Force Base colony in New Mexico did some behavioural research, albeit under the extraordinary conditions of a desert enclosure (van Hooff, 1970; Kollar, 1972). Some universities have had chimpanzee colonies, notably Oklahoma (Wallis, 1985) and Stanford (Kraemer, 1979), but of these only Central Washington's centre for sign language studies (Fouts, 1989) and Ohio State's centre for cognitive studies (Boysen & Berntson, 1989) continue to be active. No North American zoo has consistently done behavioural research on chimpanzees over the years, though the San Diego Zoological Society's colony of bonobos deserves attention (de Waal, 1986).

In Japan, the Japan Monkey Centre and the Primate Research Institute of Kyoto University, both at Inuyama, keep chimpanzees. However, only recently has behavioural research been published in English (Matsuzawa, 1985). Zoo research on chimpanzee tool-use has been in the Tama Zoological Park, Tokyo (Kitahara-Frisch & Norikoshi, 1982; Sumita *et al.*, 1985; Kitahara-Frisch *et al.*, 1987).

In Europe, few laboratories and apparently no universities keep many chimpanzees for behavioural study. Of the former, the Primate Center TNO of Rijswijk, Netherlands, keeps chimpanzees in small, indoor cages (Dienske & van Vreeswijk, 1987). However, research in well-appointed zoos is a long-standing tradition, of which Arnhem's colony is the best-known example (de Waal, 1982). Such naturalistic research followed liberalising reforms in housing and husbandry (Mottershead, 1963). The only research on tool-use in captive bonobos was done in three western European zoos: Antwerp, Frankfurt and Stuttgart (Jordan, 1982).

Methods of study

Chimpologists, like scientists studying other primates, come from several academic disciplines: anthropology, biology, psychiatry, psychology. The result in primatology is probably a more varied blend than is found in any of the other specialities devoted to a particular kind of animal. Where else might proponents of Levi-Strauss, Darwin, Freud and Piaget wrangle over the same data?

Expression of these differing viewpoints comes in the methods chosen for recording and analysing data, and their origins lie in the implicit questions to which answers are sought or, more rarely, in the explicit hypotheses which are posed for testing. Failure to take account of these differences has sometimes led to confusion, and even conflict. Below, I will deal first with gross variation in setting, such as field versus laboratory, then with the niceties of data collection and treatment, such as sampling regimes.

Studies in nature

By *field studies* is meant research on chimpanzees living in the African wilds – eating, sleeping, travelling, grooming, mating – outwith the undue influence of human beings. (Of course, the rub lies in the adjective 'undue'.) The ideal field-study would take place in a naturally bounded tract of wilderness big enough to hold a viable population of apes. It would have several communities in order for genes and habits to be exchanged through migration. The subjects would tolerate observers at close range. There would be a full array of basic resources such as air, soil and water. Co-existing with the apes would be a full range of fauna and flora, as predators, prey and competitors. Human beings would be there too, but only to the extent of being part of a self-sustaining ecosystem, for example as gatherer-hunters. At the same time, the site would be accessible to modern transport, services and communications. Everything would need to be secure for at least the average lifespan of a chimpanzee living to old age, say 40 years. Finally, enough funds would be available to last the same period. Such an Eden has not yet been found, and in the meantime chimpologists do the best they can.

In terms of time, studies of chimpanzees range from a few days (Moore, 1986) to over three decades (Goodall, 1986). So, how is one to mark out the continuum? A sensible cut-off point is between *short* studies of less than a year, and *long* studies of more than that, as the latter allow for *intra*-annual variation. Even better are studies of 2 years or more, which allow the tackling of *inter*-annual variation. Both aspects of cyclicity have proved to be important, especially in very seasonal habitats (McGrew *et al.*, 1981).

In terms of space, studies of chimpanzees vary from those that move from one bivouac to another in a survey, to those that establish a single-sited station. An example of the former was Kano's (1972) 10-month foot safari over 20 000 square kilometres of western Tanzania. An example of the latter is the Gombe Stream Wildlife Research Centre in Kakombe Valley that caters to the best-known community of apes in the world. Other studies are multi-sited but within the same area. In the

Mahale Mountains, research focusses on unit-groups at three camps: Bilenge, Kansyana and Myako. Finally, some studies seek wider comparisons. The Stirling African Primate Project has sought to apply the same methods to studying eastern (Gombe, Mahale), central-western (Belinga, Lopé) and far western (Assirik, Sapo) chimpanzees.

Another continuum is that of habituation. (Schaller, 1963, may have been the first to apply this Pavlovian term in the wilds: T. Nishida, personal communication.) Unhabituated chimpanzees avoid human beings and flee upon detecting them. This may be a legacy of past hunting or trapping, and can be quite discouraging to researchers. In 22 months of study at Lac Tumba, Horn (1980) saw bonobos only 24 times for a total of 6 hours. On the other hand, fully habituated chimpanzees go about all aspects of daily life from dawn to dusk, seemingly unconcerned by human observers only a few metres away. The record must go to Riss & Busse (1977) who followed an adult male (Figan) at Gombe for 50 days in a row, clocking up 563 hours of observation. Most subjects of study fall in between these extremes, and age, sex and individual differences are predictable. Usually, the first chimpanzees to tolerate humans at close range are adult males and the last are adult females with young offspring. This creates grounds for bias in observation, and regrettably few studies have reported systematic data on progress to shorter observation distances and longer observation times (but see Tutin & Fernandez, 1991).

With habituation comes more detailed knowledge of subjects. At first, data may be collected only in terms of age–sex classes, before individuals are recognised. When whole communities are identified, then immigration can be inferred when strangers appear. When adult females and their neonates are known, matrilineal kinship ties can be taken into account. Patrilineal kinship can be inferred if a female consorted with only one male during her oestrous cycle of conception; this entails counting back about 7.5 months from a full-term birth (Tutin, 1979). Direct calculation of relatedness through DNA fingerprinting has yet to be done in wild chimpanzees, but has proved successful in captivity (Washio *et al.*, 1989).

Another factor which usually varies with the degree of habituation is the precision of the data. The most useful data are those from unobscured and continuous observation at close range. (If these can be captured on film or tape for re-viewing, this is even better.) Such conditions are never guaranteed in the field, but are sometimes taken for granted by lucky field-workers studying fully habituated subjects. Less useful are data which are second-hand, or 'noisy', or incomplete, or otherwise only opportunistically available. This is typical of partly habituated subjects. Finally, it is possible to collect data on wholly unhabituated apes. One way is to watch them without their knowing it, from hiding. This may be necessary in a short study, but it depends on being able to find the chimpanzees. It is most efficient when the movements of the subjects can be predicted, such as their seeking out seasonally productive and patchy resources like fruiting trees (Ghiglieri, 1984) or termite mounds (McGrew & Collins, 1985).

The other way to deal with wary subjects is to use circumstantial evidence such as artefacts or other traces that are the products of chimpanzees' activity. Such indirect evidence can be useful even if the apes are *never* seen or heard, and so are especially welcome in the early stages of a study. Examples are tools, nests, feeding remains, footprints, hairs, odours, urine and faeces. Of course, the usefulness of such clues is proportional to their validity, which means that criteria for acceptance of such data must be made explicit and rigorous (McGrew *et al.*, 1988).

To the informed eye, a chimpanzee's bark tool is an unmistakable artefact, and so a clear sign of presence. However, a nest may be less useful if found in an area that also has gorillas, who also build nests. Spat-out seeds and skins may show fruit-eating, but incisor-marks may be needed to distinguish the diners from monkeys. The maxim is 'Presence of evidence shows only possibility, not certainty'. Sometimes a battery of such indicators can strengthen the case. McGrew *et al.* (1979*a*) gave eight criteria for inferring the presence of termite-fishing from artefacts alone.

Sometimes indirect evidence is even better than direct but incomplete evidence. Consider the messy subject of *faeces* (see Figure 2.4). For all known populations of wild chimpanzees, eating animal prey is uncommon. Thus, it may be missed altogether in observations of partly habituated subjects, due to sampling error, and this may produce a false-negative conclusion. But *all* kinds of animal prey known to be eaten by chimpanzees have indigestible parts, whether these be bone, teeth, hair, skin, shell, feathers, scales, wax or chitinous exoskeleton (Moreno-Black, 1978). Even soft tissues may leave traces: for example, muscles contain blood that contains haematin, which is detectable (Spencer *et al.*, 1982). Faecal analysis is important because what goes in one end of the alimentary tract *must* come out the other, minus what has been assimilated (Milton & Demment, 1989). Further, it is hard to see how

Figure 2.4. Arm of a bushbaby recovered from chimpanzee faeces, Assirik, Senegal. Scale is in centimetres.

such dietary data could be biassed across subjects, as all apes defecate daily and apparently randomly. This allows for systematic comparison across individuals, age and sex classes, groups, and populations (McGrew *et al.*, 1979*b*). Of course, faeces may yield other useful information too, such as intestinal parasites, pathogens or, through metabolised hormones, the sex and reproductive state of the depositor (Fry, 1985).

Finally, an obvious but crucial point must be put explicitly: chimpanzees do not exist in a vacuum. Everything that they do reflects the bio-physical environment in which they live. Thus, field primatologists must be ecologists, even if they are interested only in behaviour. Consider the following chain: social status is a function of health is a function of nutrition is a function of diet is a function of habitat is a function of climate and the elements. Thus to understand chimpanzee life one needs data on soils and surface water, on climate (temperature, rainfall, humidity, sunshine), and on flora and fauna (type, numbers, structure, distribution, composition). All of these vary over time, which means phenology. Perhaps needless to say, no study has yet dealt satisfactorily with all of these, but some have come closer than others.

Methodological issues

If methods are so straightforward, why are there problems? The answers lie in the nature of the subjects and of their students. The very presence of the latter affects the former (Weider, 1980). Chimpanzees are intelligent enough to exploit humans as accessories in quarrels with each other, or in response to scientific cupidity. Thus, observation in nature is really negotiation. The wild apes are volunteers, and therefore are very different from their captive counterparts.

Provisioning is a good example of a thorny issue. It is the acceptance from humans of prized items by wild chimpanzees in exchange for allowing themselves to be watched (see Figure 2.5). The idea is that fearful or skittish reactions to humans can be replaced by neutral or positive ones through repeated pairings of treats and observers. Typically provisioning means giving a high-energy food (banana, sugar-cane, pineapple, citrus fruit) at a fixed point, such as a clearing, to tempt the apes into the open for longer periods and at closer range. (A variant of this, as practised at Kasoje, is the 'moveable feast' technique, which means taking food to different places and calling in the apes by imitating their pant-hoots.) Sites where provisioning has been used are Beni, Bossou, Gombe, Kanka Sili, Kasoje and Wamba. Sites where it has not been used are Assirik, Belinga, Ishasha, Kibale, Lomako, Lopé and Tai. It is likely that the success of provisioning mostly depends on the apes' already knowing about artificial foods, as from crop-raiding (Dunnett *et al.*, 1970).

Some problems of provisioning are obvious: it is likely to alter natural patterns of feeding and ranging and, by changing the chimpanzees' energy-budget, to affect daily rhythms of activity. The extent of distortion seems proportional to the amount of feeding and the type of food given. Several critics (Reynolds, 1975) have raised more specific points such as its indirect effects on meat-eating. But however provocative the speculation, the only empirical study of the effects of provisioning remains that of Wrangham (1974). He found at Gombe that heavy artificial feeding brought chimpanzees and baboons into frequent conflict in the provisioning area. In the absence of other hard evidence, it seems best to avoid provisioning, or if it is used, to cut the amounts and periods to the minimum. Another way to achieve tameness is to give

Figure 2.5. Chimpanzees eat provisions of bananas, Gombe, Tanzania. Adult male (right) evades reach of adult female (left). (Courtesy of C. Tutin.)

treats which are less nutritionally intrusive, such as salt or even cardboard (Goodall, 1971). Once habituation has been achieved, there seems little point in continuing provisioning.

Given the problems, why provision at all? The short answer is that no study has yet succeeded in fully habituating a whole group without it, although this has almost been achieved at Tai (Boesch & Boesch, 1989).

That human intrusion can influence tool-use by wild chimpanzees is clear. The celebrated rise to alpha status by Mike, a small but smart adult male at Gombe, came through his use of empty paraffin tins as noisy accessories to enhance his charging displays (Goodall, 1971). The first recorded use of levers by wild chimpanzees came in their impatient prising open of cement and metal boxes built to contain and dispense bananas (Goodall, 1968, p. 207). The sticks used so impressively as missiles and clubs by Guinean chimpanzees in response to a stuffed leopard had been pre-cut and strewn about the observation area (J. van Orshoven, personal communication). What is often *not* clear is whether or not such intervention has altered the apes' acts outwith the provisioning site. Before-and-after comparisons are needed. This tricky issue is covered in more detail in Chapter 8, as is that of possible longer-term, unintentional human effects on apes (Eaton, 1978; Kortlandt, 1986).

Studies in captivity

Chimpanzees in captivity are deprived. No matter how spacious or stimulating their confinement, they are denied freedom of movement and thus freedom of or from association. Most are denied a lot more, and so interpreting behavioural data from captive subjects is at best a ticklish salvage operation. At worst it is a perilous exercise, both scientifically and ethically. A useful rule-of-thumb is that the closer the captive environment is to a natural one, the more valid are the data obtained. This may seem a truism, but it is rarely mentioned by laboratory scientists (for extended discussion of these points, see McGrew, 1981*b*).

This is not to say that conditions in captivity are all limited to the same degree. Far from it. Consider the following spectrum: some chimpanzees live alone in small, bare cages indoors; they eat only artificial foods, sleep on the floor, and have as objects of amusement only their excreta. Other chimpanzees live in groups whose age–sex composition is like that in the wild; they occupy large outdoor enclosures with some natural vegetation, eat a varied menu including natural foods, sleep in self-made nests, and have a rich array of objects to handle. Most chimpanzees in captivity are in conditions that fall between these two extremes. On a different continuum, some chimpanzees live in human homes, wear clothes, eat at table, and watch wildlife documentaries on television.

The problem is how to make sense of data from this variety of settings. How can findings from laboratory, zoological garden and household be compared? Put another way, how can the effects of social, intellectual, sensory, motor, and nutritional deprivation, to name just the obvious ones, be disentangled?

Actually, there are two sets of problems. One is how to avoid false-negative results, that is, under-estimating abilities because deprived conditions allow only deprived performance. A captive chimpanzee could hardly be expected to make tools unless given raw materials. The other is how to avoid false-positive results, that is, over-estimating performance as typical when really it is induced by artificial circumstances. A captive chimpanzee may use a needle and thread but we should not expect sewing in a wild counterpart. (These two sets of problems are not symmetrical. What chimpanzees *do* not do is not equal to what they *can* not do; what they *do* do cannot be denied them.)

The nature and extent of deprivation suffered by captive chimpanzees is well known to affect their use of objects. Early studies showed that prior exposure or lack of it affected performance with sticks in a simple food-retrieval task (Schiller, 1952). Remedial opportunities over only 3 days turned non-stick-users into users (Birch, 1945). Better-controlled later studies were less encouraging: deprived 2-year-olds showed little interest in objects and spent most of their time in stereotyped activities such as rocking, head-banging and eye-poking (Menzel *et al.*, 1963). Even after 4–6 years of social living outdoors with wild-born peers, the deprived chimpanzees still showed deficits in the task and some never succeeded (Menzel *et al.*, 1970). We should no more generalise about the tool-using capacities of chimpanzees on the basis of such impoverished data (Kitahara-Frisch, 1977; Tomasello *et al.*, 1987) than we would generalise about the mental abilities of human beings on the basis of data from institutionally deprived children (Lane, 1977).

Special comment is needed about a particular kind of captive chimpanzee, often described as 'semi-free-ranging', 'semi-natural' or 'naturalistic'. This usually means a group of chimpanzees living outdoors in a spacious enclosure with natural vegetation and minimal human interference. What keeps them from being a natural population? First, they are not free to disperse, being bounded, if only by a natural barrier such as water around an island. (Chimpanzees do not swim.) Second, they need supplementary feeding, if only for part of the time or for part of the diet. Examples are transported chimpanzees on a riverine island in Gabon (Hladik, 1973) or released chimpanzees on estuarine offshore islands in Liberia (Hannah & McGrew, 1991). In both cases tool-use occurred spontaneously but may have originated either in their period of captive contact with humans or in their natural upbringing before capture. Such populations may be instructive intermediates between the truly natural and more typically captive states.

Finally, a reminder for those purists dismissive of studies done in the contrived conditions of captivity: It is logically impossible to do an experiment in nature. To control for variables is to intervene, and intervention of the kind required is unnatural. To those who would do without experiments, the challenge is clear: How can one otherwise choose between alternative hypotheses, at least in a complicated creature like a chimpanzee? (With enough data and powerful-enough multi-variate analyses, this can be bypassed in some cases, but in practical terms it is daunting.) For example, it is all very well to say that social learning is involved in the individual's acquisition of tool-use in nature. But who is able to watch the development of an *asocial* chimpanzee in the wild? Such unfortunate youngsters do not survive. Social life may be a necessary condition, but to show that it is a sufficient one requires an experiment.

Collecting data

In general, for studies of chimpanzees, methods of collecting and treating data, from design of recording systems to statistical testing, are much like those in any other branch of animal behaviour. (For a recent review see Martin & Bateson, 1986.) Only special points of interest to chimpology are raised below.

Despite long-standing debate on the subject (Altmann, 1974), many field-studies of chimpanzees do not make clear their methods of *sampling*. For instance, differences between individuals are more likely to be real when focal-subject sampling is used, as scan-sampling is easily biassed towards more easily observable patterns (Martin & Bateson, 1986). Furthermore, any alternative to true frequencies and durations is subject to various biasses, and even these measures usually need to be converted to rates, if individuals are watched for differing periods (Altmann & Altmann, 1977).

In the field, subjects are rarely equally observable all the time, so periods of poor observation of varying degrees must be treated differently. For example, in 1 minute one may see that individual *A* is doing pattern *X*, in the next minute that someone (but who?) is doing *Y*, and in the next minute only that *no* one is doing *Z*.

Despite the fact that chimpanzee life is complicated and sometimes subtle, and that data are often pooled from a team of observers, there seem to be no cases of inter-observer *reliability* testing in studies of wild apes (Caro *et al.*, 1979). Even intra-observer testing is hard to find (but for an example see Plooij, 1984). However, sometimes reliability testing in captive studies can be linked to field-studies, to their mutual benefit (Kraemer, 1979).

Machlis *et al.* (1985) showed that the published literature in ethology is rife with *pooling*. This occurs when the same subject contributes more than one datum to a set. Pooling almost always leads to inflation of sample sizes, and thus to false-positive statistical significances. This is equally true of studies of chimpanzees, with its invalidating consequences (McGrew, 1979).

For data analysis, some articles contain only tabulated data but not *statistical testing* of these. In some cases, this applies to whole studies (Jones & Sabater Pí, 1971). Thus, though differences between individuals, ages, sexes, groups, etc., are claimed, there is no way to know of their validity. Thus they remain hypotheses, not findings. (Sometimes enough raw data are provided to allow readers to do the statistics, however.) Even when statistical testing is done, key details are often omitted, such as which test was used, or whether the sampling distribution was one- or two-tailed (Sabater Pí, 1979).

Finally, some authors, especially field-workers, choose to refer to data but not to present it. Thus, Sabater Pí (1979) referred to faecal specimens providing information on diet, but made no further mention of them. Others prefer to rely on qualitative description without numerical analyses. In the cases of preliminary findings (Hannah & McGrew, 1987) or telling anecdotes (Plooij, 1978), this may be all that can be said. However, if a comparative study makes a major claim, such as that savanna-living chimpanzees use weapons more than do forest-living ones (Kortlandt, 1965), then evidence is required.

Conclusion

This chapter began with the statement that chimpanzees have been studied more intensively and extensively than any other African primate. It ends on a discouraging note, with a list of methodological failings. What, then, is to be concluded? Perhaps all that can be echoed is the perennial scientist's plaint: We know a lot about chimpanzees but we also have a lot to learn. No findings, whether from field or captivity, can be taken without careful scrutiny.

3

Chimpanzees as apes

Introduction

This chapter has three aims:

(1) To compare the extent of tool-use across living apes;
(2) To relate variation in tool-use by homology to phylogenetic relationships and by analogy to variation in other features;
(3) To synthesise these findings so as to infer aspects of tool-use by ancestral hominoids that have implications for understanding the origins of material culture.

Each aim is a further step removed from the data.

The first aim entails updating of the evidence, as new findings on tools used by apes continue to mount. More and more it is clear that context is important: how an organism behaves in captivity may or may not reflect its actions in nature. Here, the exercise shows not only the state of play but also persisting gaps in knowledge. Many question-marks remain about the tool-use of even these well-studied mammals.

The two parts of the second aim present different problems. The fossil record for apes is minimal. Molecular anthropology now provides a wealth of data for tackling phylogeny, but the conclusions do not always agree (Miyamoto *et al.*, 1987). Evolutionary relationships among the African Pongidae and Hominidae remain unresolved, although a consensus seems to be emerging (Goodman *et al.*, 1990). In arguing by analogy in terms of anatomy, individual abilities and socio-ecology, the difficulty is in choosing the right variables to the right degree of specificity.

In undertaking the third aim, this chapter starts from the simple premise that each mode in the phylogenetic tree represents an ancestral hominoid. Each of these is fair game for what Tooby &

DeVore (1987) call 'strategic modelling', that is, the construction of conceptual models of human and non-human primate behaviour based on current understanding of evolutionary theory. Here, such a conceptual model is referentially based on living apes, with all of the limitations that this entails (McGrew, 1989*c*, 1991*b*).

Such a wide-ranging exercise is bound to be superficial, incomplete and frustrating. For example, *bipedal locomotion* has often been linked with tool-use in evolutionary reconstructions (Hewes, 1961), so it should be a prime candidate for examination in the apes. However, apes rarely go bipedal, and when they do it may be misleading. Gibbons on the ground always move bipedally, but being on the ground is unnatural for an arboreal creature. Wild chimpanzees being artificially fed go bipedal to carry away their booty, but they rarely show upright locomotion at other times.

Sources and methods

To survey all primary sources would be exhausting, so this effort relies upon Beck's (1980) and Tuttle's (1986) exhaustive syntheses. Other material comes from original reports, relevant reviews, or comprehensive anthologies. Despite this, there are surprising gaps in knowledge: for example, a rigorous, comparative assessment of overall intelligence in great apes is overdue, having not been done since Rumbaugh (1970). Also, perhaps more surprisingly, a systematic, empirical attempt to induce tool-use in captive lowland gorillas remains to be tried.

To contrast the living apes, they are here split into six types. These are taxonomically messy, but ecologically revealing (see Table 3.1). All three geographical races of chimpanzees are lumped, as are Bornean and Sumatran orang-utans. Gorillas are split into lowland and highland (*G. g. beringei*) forms on ecological grounds. All forms of gibbons, including siamang, are also lumped. This scheme may offend partisans (cf. Caccone & Powell, 1989), but it seems sensibly heuristic here.

It is possible to fit five of the six types into a phylogenetic tree on cladistic grounds, indicating relative degrees of closeness and thus a sequence of common ancestry. (Comparative biochemical data on the two types of gorilla are not yet available.) Sibley *et al.* (1990) and Caccone & Powell (1989) have done this on the basis of DNA–DNA hybridisation–dissociation tests. Miyamoto *et al.* (1987) and Goodman *et al.* (1990) have done so using DNA sequences at globin loci. Calibrating the 'molecular clock' to obtain absolute timing is more contentious

Table 3.1. *Taxonomy of living Hominidae (after Goodman* et al., *1990)*

Scientific name	Common name
SUBFAMILY HYLOBATINAE	Lesser apes
Hylobates spp.	Gibbons
Symphalangus syndactylus	Siamang
SUBFAMILY HOMININAE	Great apes and humans
Pongo pygmaeus	Orang-utan
P. p. pygmaeus	Bornean orang-utan
P. p. abelii	Sumatran orang-utan
Gorilla gorilla	Gorilla
G. g. gorilla	Western lowland gorilla
G. g. graueri	Eastern lowland gorilla
G. g. beringei	Highland gorilla
Pan paniscus	Bonobo (pygmy chimpanzee)
Pan troglodytes	Chimpanzee
P. t. troglodytes	Central western chimpanzee
P. t. schweinfurthii	Eastern chimpanzee
P. t. verus	Far western chimpanzee
Homo s. sapiens	Modern human beings

(see Foley, 1987*a*, for essentials), but the emerging consensus seems to be that humans and chimpanzees emerged about 6–8 million years ago. This yields a basis for comparison of tool-use across hominoids on grounds of homology.

The six types of apes are then contrasted on two fronts: use of tools (see Table 3.2), and socio-ecology, brain, hands, and mind (see Tables 3.5–3.7). The resulting classifications are crude and are useful only for qualitative comparison. 'Captive' denotes apes living in confinement where both the means and ends of tool-use are largely absent, while 'free-ranging' means the unenclosed opposite, where there is, for example, access to natural vegetation. In captivity, tool-use is termed either 'spontaneous', that is, unprompted by humans, or 'induced' by human intervention, often in a structured, experimental way. In free-ranging, the sub-division is between 'natural', which means pristine and 'human-influenced', in which wild apes are provisioned or previously captive apes are released. Six types of ape times four types of setting yields a matrix of 24 cells, each of which has a forced choice coding of one of five types. Overall, presence is probably more accurate than absence, given the whimsy of negative evidence.

For socio-ecology, in Table 3.3, the matrix has 30 cells: six types of

Table 3.2. *Use of tools by apes (and capuchin monkeys) in four settings*

	Captive		Free-ranging	
	Spontaneous	Induced	Human-influenced	Natural
Chimpanzee	++	++	++	++
Orang-utan	++	++	++	+
Bonobo	++	?	+	−
Gorilla	+	+	+	−−
Gibbon	+	?	+	+
Capuchin	++	++	?	+

++, well known from several individuals in several populations; +, recorded at least once somewhere; −−, notably absent from long-term studies of several populations; −, not seen but few or short studies, or few subjects; ?, not yet properly studied.

Table 3.3. *Selected aspects of socio-ecology of living apes*

		Diet		Social structure[d]	
Type	Vertical distribution[a]	Plant[b]	Animal[c]	Parties	Network
Chimpanzee	Ter/Arb[b]	Frug	M, B, E, I	Variable	Closed
Bonobo	Arb/Ter	Frug/Fol	M, E, I	Variable	Closed (?)
Highland gorilla	Ter	Fol	(I)	Stable	Closed
Lowland gorilla	Ter/Arb	Fol/Frug	I	Stable (?)	Closed (?)
Orang-utan	Arb	Frug	M, B, E, I	Solitary	??
Gibbon	Arb	Frug	I	Stable	Closed

[a] Ter, terrestrial; Arb, arboreal.
[b] Frug, frugivorous; Fol, folivorous.
[c] M, mammal; B, bird; E, egg; I, insect.
[d] Modified from Wrangham (1987).

ape times five types of social or environmental categorisation. 'Vertical distribution' refers to whether apes spend more of their waking hours on the ground (terrestrial) or above it in vegetation (arboreal). The 'plant' portion of the diet is classified by the predominant part eaten: fruit or foliage. For the 'animal' portion of the diet the types of prey are given, such as mammal. For social structure Wrangham's (1987) criteria for membership of parties are followed whenever possible: 'stable' means that composition of the party stays constant over months, while 'variable' means that it changes over weeks, days or hours. For social net-

Table 3.4. *Living apes ranked in terms of tool-use, phylogeny and features of socio-ecology (see text for details)*

	Tool-use	Related to human	Terrestriality	Faunivory
Most	Chimpanzee	? Chimpanzee	Highland gorilla	Chimpanzee
	Orang-utan	Bonobo	Lowland gorilla	Orang-utan
	Bonobo	? Highland gorilla	Chimpanzee	Bonobo
	Lowland gorilla	Lowland gorilla	Bonobo	Gibbon
	Highland gorilla	Orang-utan	Orang-utan	Lowland gorilla
Least	Gibbon	Gibbon	Gibbon	Highland gorilla

works, 'closed' means that social relations are mostly confined to members of the unit, to which entry by outsiders is resisted. An 'open' network is one that is not closed.

Features of brain, hands and mind (Tables 3.4 to 3.7) were chosen based on an arbitrary threshold of data being available from at least four of the six types of ape. Almost all studies were bedevilled by small samples and varying methods of data collection and analysis. Accordingly, results were treated only as ordinal data, that is, capable only of being ranked.

Patterns of tool-use

Chimpanzee

Chimpanzees in all settings use tools regularly (Beck, 1980; Tuttle, 1986). In the wild they use a variety of tools made from a variety of materials to accomplish a variety of tasks. This is true of the eastern, central-western and far western geographical races, and is known in habitats ranging from savanna to evergreen forest. Well-known types of tools include probes of vegetation to obtain social insects (Nishida, 1973), hammers of stone to crack open nuts (Boesch, 1978), sponges of leaves to soak up fluids (Goodall, 1968), and weapons of woody branches to deter predators or to dominate opponents (Kortlandt, 1965). Wild chimpanzees also *make* tools and show flexibility in doing so: they use a variety of raw materials to make the same tool, such as twig, vine or bark to fashion a probe for termite-fishing (Goodall, 1964). Also, they use the same raw material to make various tools: for instance, a leaf may be modified to be a sponge, napkin, probe or billet-doux (Nishida, 1980*b*).

Marked contrasts occur across populations (see Chapter 7), and some of the differences seem to be cultural, resulting from social traditions and reflecting more than just environmental affordances (McGrew *et al.*, 1979*a*). Finally, some groups of wild chimpanzees seem to have more impressive tool-kits than others (Boesch & Boesch, 1990), but it is not yet clear whether these contrasts are real or are artefacts of differing observational conditions or techniques.

A similar variety of tool-use is shown by free-ranging chimpanzees influenced by varying degrees of human contact. Chimpanzees released after years of confinement onto forested islands use hammers to crack open nuts (Hannah & McGrew, 1987). Crop-raiding chimpanzees in a relict Guinean population living near human settlement showed similar use of hammer-stones (Sugiyama & Koman, 1979). Both before and after provisioning, chimpanzees at Gombe showed the same kind of tool-use directed to natural prey, but they also added new tool-use to their repertoire, such as using levers to prise open metal boxes that supplied bananas (Goodall, 1968).

In captivity, chimpanzees spontaneously show every mode of tool-use seen in the wild (Beck, 1980), with the degree and range of expression being largely a function of opportunity (see Figure 3.1). For example, given an artificial 'termite mound' containing prized food, chimpanzees at Edinburgh Zoo made tools from their bedding branches to probe for it (Nash, 1982). Further, captive chimpanzees invented new types of tools to solve new problems, such as poles used as ladders to escape from enclosures (Menzel, 1973) (see Figure 3.2). The most extensive tool-users are chimpanzees reared in human homes, who learn by imitation to use many household implements, from door-key to fishing rod (Temerlin, 1975).

Studies of induced tool-use in captive chimpanzees began almost 80 years ago, with the early efforts of Kohts (1935) and Köhler (1927). Köhler set standards with tasks which are still used: a rake to obtain an out-of-reach incentive, boxes stacked to obtain an incentive suspended overhead, etc. Circumstances that induce tool-use vary from merely providing materials in a structured setting, such as crayons and paper for drawing (Smith, 1973), to carefully demonstrating how to solve problems (Hayes & Hayes, 1954). Tool-use has since been linked to other intellectual tasks, such as one chimpanzee's use of a symbol-system to

Figure 3.1. Spontaneous uses of a tyre by an adult male (Shadow), Delta Primate Center, USA: (*a*) chaise-longue for rest; (*b*) toy for solitary play; (*c*) container for carrying water; (*d*) support for sexual inspection.

ask for a tool from another, with the recipient using the tool to obtain a food-item that both then share (Savage-Rumbaugh *et al.*, 1978).

All in all, chimpanzees impressively perform tool-making and tool-use in all settings. However, this reassuring uniformity does not apply to other apes.

Figure 3.1. *cont.*

Bonobo

Only recently have field-studies of bonobos reached the stage at which good behavioural observations have accumulated (see chapters in Susman, 1984; White, 1992). Primatologists at both sites in north-central Zaïre (Lomako and Wamba) have reported no evidence of habitual tool-use, although it has been keenly sought. The only established tool-use seen so far has been five cases of leafy twigs being used as partial shelter from rain (Kano, 1982). This similar absence at both sites holds

Figure 3.2. Ladder-use to escape from an outdoor enclosure, Delta Primate Center, USA: (*a*) erecting the plank; (*b*) climbing the plank; (*c*) balancing atop the plank.

though the methods of study differ: at Wamba the apes are heavily provisioned at an artificially cleared feeding site, while at Lomako no provisioning is done.

In captivity no studies of induced tool-use seem to have been tried, perhaps because few bonobos are in captivity and even fewer of them are in laboratories. However, Jordan (1982) reported observations of groups in zoological gardens in western Europe. Using Beck's (1980) modes of tool-use, she reported a range of spontaneous behavioural patterns indistinguishable from that of chimpanzees.

Orang-utan

Orang-utans present a puzzle. Long-term field-studies, both at Tanjung Puting in Borneo (Galdikas, 1982) and at Ketambe in Sumatra (Rijksen, 1978), have yielded mostly negative results. Only Tanjung Puting's orang-utans have shown limited technical inclinations: dropping twigs and toppling snags from the canopy in agonistic displays were the most common. Rubbing the face with a handful of leaves was the most enigmatic (Galdikas, 1982, 1989). Most conspicuously absent were cases of tool-use in feeding; these findings after years of careful study confirm those from medium-length studies at other sites (MacKinnon, 1974; Rodman, 1977). (Chevalier-Skolnikoff *et al.* (1982) have argued that orang-utans in nature use tools more often than usually credited; however their definitions are broader than Beck's (1980).)

No intentional provisioning of wild orang-utans seems to have been done. However, both Tanjung Puting and Ketambe doubled as rehabilitation centres as well as field-sites, so provisioning has happened accidentally (Rijksen, 1978, p. 369). Captive apes such as those confiscated by wildlife officials were thus released into the same habitats in which their wild counterparts lived. These released orang-utans showed a rich array of tool-use, some of it remarkably inventive, such as use of floating objects to raft across a river (Galdikas, 1982; Russon & Galdikas, 1992). Much of their tool-use involved artificial objects and tasks, but other usages were naturalistic, such as trying to open spiny fruits with a stick (Rijksen, 1978, p. 84).

Because of being studied largely in zoological gardens and not in laboratories, orang-utans in captivity are unparalleled tool-users (Lethmate, 1982). For example, one orang-utan wound 'wood-wool' around a cracked stick to mend it for use as a tool. Such 'creative' conjunction (see Table 6.4) in tool-making has not been seen in any other non-

(a)

(b)

(c)

human species (see Figure 3.3). Lethmate made a detailed comparison of orang-utans and chimpanzees on 23 categories of tool-use and five of tool-making. There were few differences. These accomplishments apply equally to spontaneous and to induced tool-use. Perhaps the most striking example of the latter is that of the making and using of flaked stone tools (Wright, 1972). After only a few hours of human demonstration, the ape used a quartzite hammer to knap flint flakes, the sharp edges of which he used to cut a cord, allowing access to food in a box.

Highland gorilla

Highland gorillas also pose problems. In the wild, the negative evidence of tool-use is overwhelming. Many years of study in the Virunga Volcanoes of Rwanda, Uganda, and Zaïre have produced no signs of tool-use (Fossey & Harcourt, 1977; Watts, 1984). This holds despite day-long observations of hand-use by relaxed subjects at a few metres' distance on the ground (Byrne & Byrne, 1991). Highland gorillas have been studied only in almost pristine conditions, without rehabilitation, provisioning or crop-raiding, so nothing can yet be said about human-influenced tool-use in free-ranging animals.

In captivity, there are now no highland gorillas to be studied, but there were two early investigations. Yerkes (1927) thoroughly tested and re-tested a young female on a battery of Köhler-like tests. Her performance improved somewhat with age, but Yerkes remained disappointed. Carpenter (1937) watched a pair of males in the San Diego Zoo for about 6 weeks. Carpenter's observations were casual, but he saw spontaneous use of containers for drinking.

Lowland gorilla

Lowland gorillas are even more problematical. In the wild, the western form has scarcely been studied behaviourally, with the best data coming from the eastern gorillas of Kahuzi-Biega in Zaïre (Goodall, 1979). No tool-use has been seen. Further, no indirect data such as discarded tools have been found, even in areas where sympatric chimpanzees leave such

---

Figure 3.3. Male orang-utan induced to make and to use a compound tool of segments of pipe and stick: (*a*) combining elements; (*b*) using short version to get out-of-reach object; (*c*) using long version to get suspended object. (Courtesy of J. Lethmate.)

circumstantial evidence (Jones & Sabater Pí, 1971; Tutin *et al.*, 1991). Possible human influence on free-ranging lowland gorillas remains to be assessed. No wild population has yet been provisioned nor has rehabilitation into the wild of captive gorillas been tried.

In captivity, there are only scattered reports in print of spontaneous tool-use: Wood (1984) reported that gorillas in a large captive colony modified branches into sticks in order to rake in food lying beyond reach outside their cage. This habit was well established in the captive colony that provides the most stimulating social and physical environment yet devised; this probably says something about the socio-ecologically impoverished conditions in which most gorillas are kept. Surprisingly, no systematic studies of induced tool-use seem to haved been tried in lowland gorillas, except for bits and pieces with infants (Gomez, 1988; Natale *et al.*,1988; Parker, 1969). This is puzzling, given the equivalent opportunities for testing to those for orang-utans in zoological gardens. There have been studies of objects manipulated in intelligence testing, such as patterned-string problems (Fischer & Kitchener, 1965) or Piagetian research (Redshaw, 1978), but these also seem to have been limited to youngsters.

Gibbon

Field-studies of gibbons pre-date those of all other apes, having been done for over 50 years in various parts of south-east Asia (Whitten, 1982). None of these studies has reported tool-use, apart from the occasional dropping of branches onto observers below. However, all of the findings have suffered from the limiting conditions of observers on the ground watching subjects high in the trees. The studies of free-ranging gibbons loosed onto islands have been brief or intermittent, and superficial. Only one case of tool-use seems to have been seen: Baldwin & Teleki (1976) saw a young female repeatedly use a leaf as a sponge to dip water from a pool. A release of captive gibbons into the wild in Thailand yielded few behavioural data and no mention of tool-use (Tingpalapong *et al.*, 1981). Provisioning of wild gibbons remains to be done.

In captivity, only one gibbon's and one siamang's spontaneous tool-use have been described. Rumbaugh (1970) reported how a young female gibbon used a cloth as a sponge and a rope to make a swing, and a young siamang also did the latter (Anonymous, 1971). Surprisingly, there seem to have been no systematic attempts to elicit tool-use from captive gibbons. Perhaps investigators have been deterred by the unpromising anatomy of the gibbon's hands as manipulatory organs. However, Beck (1967) showed that although gibbons failed a string-pulling task when it was presented on a flat surface, they quickly passed if the strings were elevated for easier handling.

In summary, living apes show a wide variety of tool-use, from chimpanzees with their frequent and diverse instrumentation in all settings to gibbons with their total of two anecdotes. If ranking the six types of ape seems premature, then they can be nominally split for analysis into three tool-users (chimpanzee, bonobo, orang-utan) and three non-tool-users (lowland gorilla, highland gorilla, gibbon).

Socio-ecology

If tool-use depends on setting, then environmental variables should be revealing (see Table 3.3). The six types of apes can be ranked in terms of degree of terrestriality or vertical distribution (Tuttle, 1986). This seems worth doing, as most tool-use in nature takes place on the ground. However use of a leafy sponge to sop up drinking water from tree-holes seems to be obligatorily arboreal (Goodall, 1968; McGrew, 1977), as does ant-fishing (Nishida, 1973; Nishida & Hiraiwa, 1982). Perhaps more precarious is the arboreal use of hammer-stones to open nuts, in which a bough serves as an anvil (Boesch & Boesch, 1984*b*). Table 3.3 shows the apes ranked in terms of time during waking hours spent on the ground. The range is wide, from gibbons who apparently never descend to the ground, to adult male highland gorillas who may never leave it.

All apes are primarily frugivorous, except for the high-altitude gorillas who have become secondarily folivorous in a largely fruitless environment (Watts, 1984). Ranking the six types of ape is difficult, however, as some within-species differences are greater than some across-species ones (Williamson *et al.*, 1990; Tutin *et al.*, 1991). More to the point, the plant portions of the diets of the apes cannot readily be ranked on a single variable such as nutritive value (see Figure 3.4).

Even using a criterion such as difficulty of processing is tricky: chimpanzees in Ivory Coast use hammers to crack open *Detarium* nuts in order to eat the kernels (Boesch & Boesch, 1983), but lowland gorillas in Gabon apparently break them open with their teeth (E. Williamson, personal communication). There are derived ways of rank-

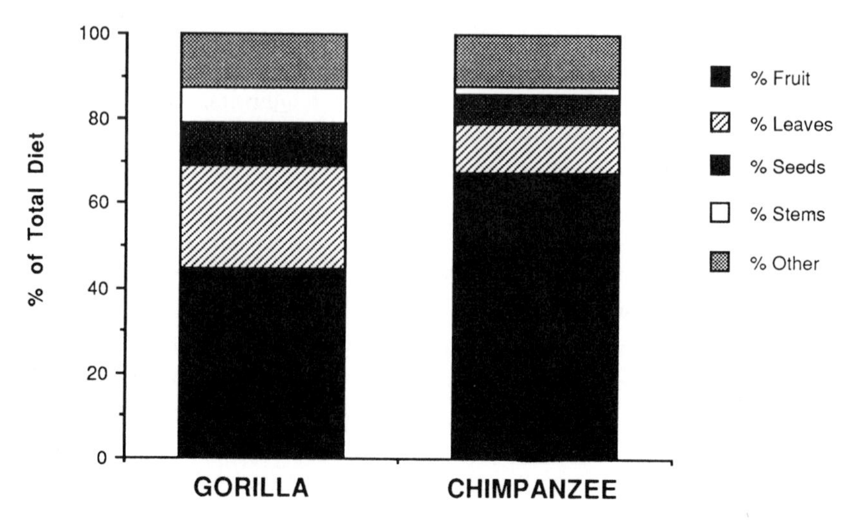

Figure 3.4 Comparison of the diet of sympatric chimpanzees and gorillas, Lopé, Gabon. Histograms show percentage breakdown by food-type. (Courtesy of C. Tutin.)

ing diet, such as in terms of quality, but that is outwith the scope of this chapter.

Animal matter in the diet, or faunivory, can be more easily ranked across the six types of ape, from most to least (Tuttle, 1986). Again, the range is wide (see Tables 3.3 and 3.4). At one end, chimpanzees in all types of habitat prey on mammals, birds and their eggs, and social insects. They use a variety of tactics geared to the vulnerability of the prey and often outcompete other sympatric predators. For example, baboons are limited to grabbing the emerging winged forms of termites, but chimpanzees use tools to extract the underground castes (Beck, 1974). Lowland gorillas in north-eastern Gabon regularly eat termites too, but these are caught by destroying their mounds and picking up the prey by hand (Tutin & Fernandez, 1983). There is but a single known case of meat-eating by a wild orang-utan: a female devoured a gibbon (Sugardjito & Nurhuda, 1981). At the other end of the scale, highland gorillas eat animal matter only inadvertently in the form of invertebrates living in the plants eaten (Harcourt & Harcourt, 1984; Watts, 1989).

Social structure varies more widely across apes than the two categories in Table 3.3 indicate (see Wrangham, 1986, 1987). The problem here is how to rank sensibly on a single scale a multi-dimensional phenomenon? Attempting such ranking with the two listed variables illustrates this: chimpanzee parties may be less stable than those of bonobos, but how does one compare either of these with the medium-term constancy of the harem in gorillas or with the longer-term fidelity of the nuclear family in gibbons? How does one classify a solitary form like an adult male orang-utan? Is his social life infinitely stable or variable? Finally, the composition and durability of parties of lowland gorillas are yet unknown.

Similar problems bedevil making sense of social networks in the apes. At first glance there seems to be little variation, with most groups seeming to be closed. (Again, the solitary orang-utan poses a problem. What is the social unit, if any, to classify?) Gibbon families maintain unified social integrity, but the other apes show consistent sex differences characterised by female dispersal and male philopatry. Even if the two sexes could be ranked separately on a single dimension of 'openness', it is likely that the males' variation is a function of access to and competition over females, while the females' variation is a function of resources related to reproduction (Wrangham, 1986, 1987).

Overall, the apes present a diverse radiation on a variety of socio-ecological criteria. This is clear even from the small set of features dealt with here, the number of which could easily be doubled or trebled, or made more specific. No univariate analysis will do justice to this variety, and the data needed for multi-variate analysis are not yet available.

Brain

If variation in behaviour reflects variation in 'hardware', then one might expect to see variation across the apes in brains and hands. Testing this turns out to be easier in principle than in practice, as little is known about either for highland gorillas or bonobos. Few studies cover the whole range of apes. Problems of allometry loom (see Table 3.5).

Comparing brain-sizes is nonsensical without taking account of body-size, hence Jerison's (1973) encephalisation quotient (EQ) that allows the derivation of a relative measure. (Others have tried similar analyses (e.g. Stephan & Andy, 1969; Clutton-Brock & Harvey, 1980), but these do not permit comparisons across the forms of apes used here.) The higher the EQ the brainier the subject relative to the 'average mammal'. Comparing EQs, chimpanzees come top, but surprisingly gibbons come second, ahead of orang-utans and gorillas (see Table 3.5). This hardly reflects tool-use. Even more surprising is the ranking produced

Table 3.5. *Living apes ranked in terms of tool-use and various features of brains (see text for details)*

	Tool-use	Encephalisation quotient[a]	Asymmetry of structure[b]	Asymmetry of structure[c]
Most	Chimpanzee	1	3	1.5
	Orang-utan	3	4	3
	Bonobo	?	2	?
	Lowland gorilla	4	1	1.5
	Highland gorilla		?	?
Least	Gibbon	2	?	4

[a] Jerison (1973), table 16.3.
[b] Holloway & de la Coste-Lareymondie (1982), tables 3–7.
[c] Lemay (1976), table 5.

by Jerison's (p. 80) measure of 'extra' neurones (nc), which shows non-tool-using gorillas to have progressed the most beyond the level required by increasing body-size.

Given the asymmetry in structure and corresponding laterality of function in the human brain, similar relationships have been sought in apes (MacNeilage *et al.*, 1987). The data are so heterogeneous and sparse, especially in the tasks chosen for study, that meaningful comparison seems impossible, at least across the minimum of four types of ape (Marchant & McGrew, 1991). Tool-use by wild chimpanzees (Boesch, 1991) and food-processing by highland gorillas (Byrne & Byrne, 1991), seems to be equally lateralised on an individual basis, but comparable data on other apes in nature are lacking.

Asymmetry of structure is easier to compare, but the results turn out to depend on the methods and measures chosen. Holloway & de la Coste-Lareymondie (1982) compared latex endocasts of crania across sizeable numbers of four types of ape, and focussed on five measures, of which two were composites. Lowland gorillas (with the lowest EQ) showed the greatest asymmetry overall, while the paramount tool-users, chimpanzees, ranked only the third. Lemay (1976) compared smaller numbers of four types of ape on cerebral asymmetry using scaled photographs of preserved brains. She used four measures, one of which, occipito-petalia, was in common with Holloway & de la Coste-Lareymondie. On Lemay's measures, gorillas ranked joint highest overall with chimpanzees. Intriguingly, the same highland gorillas who showed laterality of functioning in chest-beating (Schaller, 1963) showed significant cranial asymmetry (Groves & Humphrey, 1973).

Table 3.6. *Living apes ranked in terms of tool-use and various features of hands (see text for details)*

	Tool-use	Thumb opposability[a]	Hand length[a]	Curvature of proximal phalange[b]
Most	Chimpanzee	42	30	42
	Orang-utan	39	28	63
	Bonobo	?	?	44
	Lowland gorilla	48	25	37
	Highland gorilla	?	?	?
Least	Gibbon	47	26	?

[a] Napier & Napier (1967), pp. 401–2.
[b] Susman (1988), fig. 3.

Hands

Given that most tool-use by apes is done with the hands, their design and dimensions must be important (see Table 3.6). Napier & Napier (1967) derived the five indices for comparison of hands across species of primates: for example, the *thumb opposability index* is the ratio of thumb length to index finger length, so that the higher the figure, the more opposable the thumb. Sadly, data for the bonobo were lacking for all of the Napiers' analyses, and later studies suggested that their scheme was simplistic: most grips used by human stone-tool-makers are neither purely precision nor purely power but a combination of the two (Marzke & Shackley, 1986).

However, bonobos were included in a recent analysis specifically designed to infer tool-use from hand structure. Susman (1988) reported that short, straight proximal phalanges in hominids are indicative of well-developed precision grips. Thus the lower the included angle of the bone, the more likely is sophisticated tool-use. By this measure, gorillas would be the most likely tool-users among the apes, and orang-utans the least likely.

Mind

If comparing apes on the grounds of 'hardware' is difficult, then comparing their 'software' is doubly daunting. With the demise of comparative psychology, it is hard to find any measure of mind (or 'intelligence') for which more than four types of ape can be contrasted. This is largely because bonobos and highland gorillas are not available for testing in

Table 3.7. *Living apes ranked in terms of tool-use and various features of mind (see text for details)*

	Tool-use	String-pulling[a]	Transfer index[a]	Self-recognition[b]
Most	Chimpanzee	2	3	++
	Orang-utan	1	1.5	+
	Bonobo	?	?	?
	Lowland gorilla	3	1.5	--
	Highland gorilla	?	?	?
Least	Gibbon	4	4	–

[a] Meador *et al.* (1987).
[b] Gallup (1987).

laboratories. Only chimpanzees have been thoroughly tested in large numbers at several places.

Four types of ape can be compared on a traditional measure: solution of patterned string problems (see Table 3.7). In these the subject must pull in one or two or more strings to which a bait is attached, though the lay-out of the correct string may be indirect or even misleading. In their massive review of primate cognition, Meador *et al.* (1987) collated findings that date back almost 50 years. Orang-utans performed best and gibbons worst (cf. Beck, 1967).

A more sensitive measure of cognitive capacity, especially for quantitative comparisons across widely separated taxa, is Rumbaugh's Transfer Index (Meador *et al.*, 1987). This standardising measure taps a subject's ability to detect a reversal of cue values in a two-choice discrimination problem. How quickly the subject 'catches on' is a useful indicator of intelligence. Gorillas and orang-utans averaged slightly (but not significantly) higher scores than did chimpanzees, and only gibbons were firmly below them.

A most elegant index of intellectual capacity is the ability to recognise oneself in a mirror (Gallup, 1987). If self-recognition signals self-awareness or even self-concept, then this is a profound capacity, even if the data are only nominal, that is, either success or failure at the task. Chimpanzees and orang-utans do recognise themselves, and lowland gorillas apparently do not (Lethmate & Dücker, 1973; Suarez & Gallup, 1981; but see Patterson, 1986). Again, the data on bonobos and highland gorillas are eagerly awaited.

Apes and their tools

Several general points emerge from this survey of tool-use. First, the data remain incomplete. Of the six types of ape, only chimpanzees and orang-utans are well-enough studied in all four settings even to begin to draw conclusions. Second, there is a disappointing lack of agreement across settings, that is, for a given type of ape the occurrence and extent of tool-use in one setting are not necessarily found in another. The best set of data comes from free-ranging apes in natural settings, but it is also the most negative. The next best is of spontaneous tool-use in captivity, and it is the most positive! Adding data from the other two, sparser settings adds little to the picture. So, what is to be made of the contrasts, especially the striking discrepancy between orang-utans in nature and their counterparts in other settings? Or, between captive and free-ranging bonobos?

Several caveats need stating. First, spontaneous tool-use is not the same as being unaffected by human influence. Spontaneous only means *untaught*, and not all 'teaching' is obvious. Human reinforcement that shapes ape learning can be given unintentionally (Galdikas, 1982; Tomasello, 1990). Second, many cases of tool-use by apes, especially young ones, may be direct imitation of foster, human caretakers (Hayes & Hayes, 1952). Such parroting of patterns acquired through strong emotional attachment to a surrogate parent may be spontaneous but inadvertent by-products which tell us little about natural adaptation. Third, older captive apes have lots of 'free' time in settings which may present a wealth of objects or, conversely, may focus their attention on a few. That bored apes incorporate these objects into daily life is thus not surprising. Yet, however artificial and contrived the captive environment, one must still explain the abilities manifest there.

The most parsimonious interpretation for Table 3.2 is that the chimpanzee is the only true tool-user, given its consistency across all four settings. All tool-use by other apes can then be written off as freak accidents or as somehow prompted by contact with human beings. If so, then many or eventually all of the question-marks in the non-natural settings will probably change to plusses as data build up. For the purposes of the present comparison, however, Table 3.4 gives cautious rankings of apes in terms of tool-use, based on Table 3.2. So, how do rankings of tool-use performance correlate with rankings on other variables?

Socio-ecologically, the results range from the puzzling to the perverse

(McGrew, 1989*c*). In terms of vertical distribution, the least technical apes are at the extremes: the arboreal gibbon and the terrestrial gorilla. The most frequent tool-user, the chimpanzee, falls in the middle, as having its day-time activity most balanced between life in the trees and on the ground.

As regards diet, there is a marked contrast between herbivory and faunivory when tool-use is considered. Plant foods do not lend themselves to easy ranking, at least in terms of the crude criteria given here. There is no obvious relation between, for example, degree of frugivory and frequency of tool-use. A possibility worth exploring is that of ranking food quality in optimal foraging terms, such as net energy gain after taking into account searching time, handling time, etc. For animal foods, however, a striking correlation appears: *The more types of animals are eaten, the more tool-use is shown.* The link between vertebrate prey and frequent tool-use is especially intriguing, given the prominence given to stone tool-use and to hunting or scavenging in palaeo-anthropology (Shipman, 1986; Toth & Schick, 1986).

With no obvious single-dimensional ranking of social structure, all that can be said here is that these data suggest no simple relation between tool-use and social life. Perhaps more detailed comparisons along the lines of Wrangham's (1987) set of 14 socio-ecological variables would yield useful results. Another approach, even finer-grained, would be to compare technological and socio-ecological variation *within* a species, when enough behavioural data become available on enough populations studied in a sufficiently similar way (see Chapter 7).

For brains, no correlations are apparent. Clearly, brain-to-body-size ratios are not correlated with tool-use. Findings on laterality of function are too messy to be helpful, and those on asymmetry of brain structure are equally disappointing.

For hands, the results across four types of ape are paradoxical: two tool-users (chimpanzee and orang-utan) have lower opposability indices than do two non-tool-users (lowland gorilla and gibbon)! The best fit to the Napiers' measures is the hand length index (i.e. hand length divided by arm length), which seems improbable. For Susman's (1988) analysis of the curvature of proximal phalanges, the results look equally paradoxical (see also Trinkaus & Long, 1990). Non-tool-using gorillas have the lowest index, not the highest!

For minds, the results from the patterned string-pulling problems match those of tool-use tolerably well, with the gorilla markedly consigned to third place behind the tool-using chimpanzee and orang-utan.

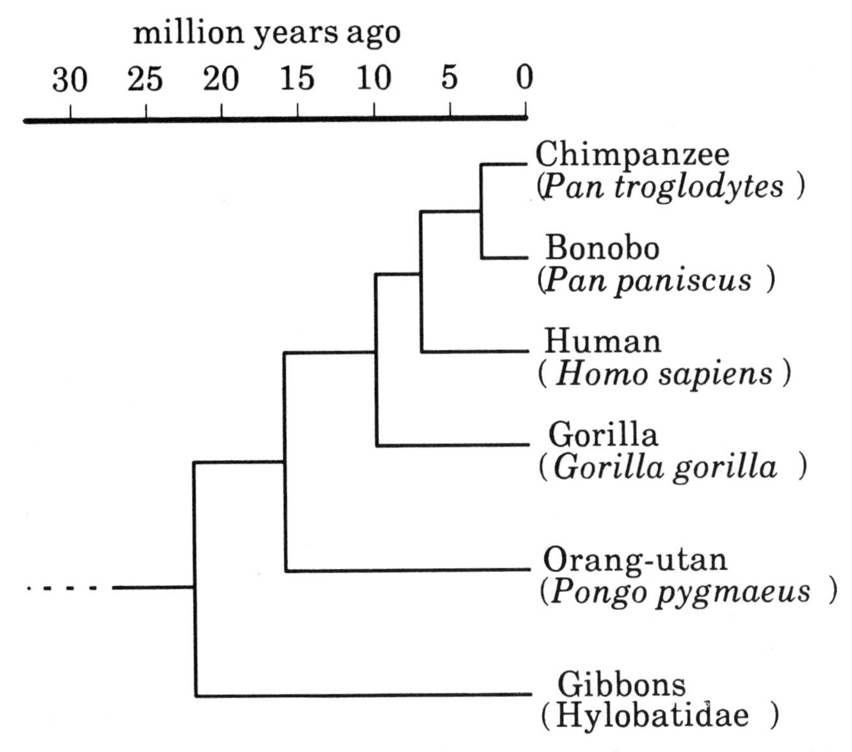

Figure 3.5. Phylogeny of living Hominoidea. Branching points indicate common ancestor of forms to the right.

However, as string-pulling *is* tool-use by some definitions, this result is not surprising! Rumbaugh's Transfer Index does not predict tool-use. Surprisingly the most congruent of the anatomical and intellectual indicators is self-recognition (McGrew, 1992*a*). Tool-using chimpanzees and orang-utans recognise their mirror-images while lowland gorillas and gibbons do not. Given that the experimental paradigm entails *no* object manipulation, the fit is even more striking (G. Gallup, personal communication.) On the basis of tool-use, it seems likely that bonobos given mirrors will recognise themselves but that highland gorillas will not.

Finally, it is possible that tool-use by apes merely reflects phylogenetic distance from humans, the supremely technological primate. To test this, it is possible to fit five of the six types of ape into a phylogenetic tree as in Figure 3.5. (Molecular biological data on highland gorillas remain to be obtained.) The data on DNA–DNA hybridisation (Caccone & Powell, 1989) and DNA sequencing (Goodman *et al.*, 1990) agree. No correlation is apparent in the living apes between extent of tool-use and phylogeny. Orang-utans are only distantly related to humans, yet are second only to chimpanzees as tool-users (see Table 3.4).

Tool-use in apes is unlikely to have been evolutionarily selected for directly, nor is it likely to be simply and univariately related to any other trait, nor yet to be some optimal mélange of being long-handed, meat-eating and self-recognising. Instead it is likely to be a by-product of a general ability for problem-solving that is expressed or not according to a set of environmental demands. Thus, gorillas are likely to be 'under-achievers' and orang-utans to be 'over-achievers' not because of *who* they are but because of *where* they are.

Ancestral hominoids

If living hominoids are a muddle, what can possibly be said about extinct ones? Complications abound. Caution is advisable, for several reasons: First, current opinion repeatedly stresses the pitfalls of referential modelling based on a single species (Tooby & DeVore, 1987). Second, by definition, living species cannot be ancestral. In chimpanzees at least, cultural evolution is likely to have occurred in parallel with, and, therefore, inseparably from organic evolution for millions of years. Third, the palaeo-archaeological and palaeontological records are biassed against perishable tools and soft food-items, so inferences rely on incomplete data.

Given these and other obstacles, a tentative, conservative model for testing hypotheses about the evolutionary origins of tool-use would be a Miocene hominoid descended from a dryopithecine (gibbon-like) ape. This ancestral proto-pongid may have had the intellectual capacities of living great apes, and its tool-use may have been more or less developed according to some local combination of socio-ecological forces. Sometimes this ape may have acted like a chimpanzee, sometimes like a gorilla, sometimes like an orang-utan, depending on where it lived. By this line of reasoning, living orang-utans are non-tool-users in nature because they have 'given up' its phenotypic expression by making a latter-day commitment to arboreality (cf. Galdikas, 1982; Lethmate, 1982). Highland gorillas have similarly traded off tool-use for making a living in a high-altitude 'salad-bowl' – that is, technology has been sacrificed for a dependable and abundant, but low-quality, folivorous diet. Paradoxically, bonobos have ended up having it both ways (or neither?), as a sort of ecological hybrid between the arboreal, frugivorous orang-utan and the terrestrial, folivorous highland gorilla (cf. Malenky & Stiles, 1991).

This model is attractive because it accounts phylogenetically and socio-ecologically for the contrast between tool-use shown by apes in captivity and in nature. *It implies that all great apes are smart enough to use tools but that they do so only in useful circumstances.* This model also leads to testable predictions: if tool-use occurs in wild lowland gorillas, it is likely to be shown by these living allopatrically with chimpanzees. Within a species, extent of the tool-kit is likely to be positively correlated with range of animal foods in the diet. Gorilla tool-use if it occurs should vary inversely with altitude, or some co-variant of it.

Another model hominoid would be an African Pliocene form ancestral to living *Pan* and *Homo*. This later-living form is perhaps more radical, as such a form need not have been a tool-user. In this scenario, the bonobo of all living apes remains least changed from the predecessor (cf. Zihlman *et al.*, 1978), with its greater reliance on terrestrial herbaceous vegetation and lack of competition from sympatric apes. (See Wrangham (1986) for the source of this line of argument. Further, bonobos may have lacked competition even from humans until recently (see Hart & Hart, 1986).) *This model implies that sometime later in antiquity, proto-chimpanzees and proto-hominids convergently invented tool-use in their respectively less and more open habitats. Even as recently as 1·5 million years ago, long after the divergence in brain size between pongids and hominids, their archaeological records could have been indistinguishable (Wynn, 1981; Wynn & McGrew, 1989).* This scenario is strikingly similar to that reported for Swartkrans cave at 1·7 –1·0 million years ago by Brain *et al.* (1988). There, they say, *Homo erectus* and *Australopithecus robustus* were sympatric, and both used (at least) stone and bone tools.

This model also produces specific predictions: In the right captive circumstances, chimpanzees should prove capable of making Oldowan-type tools. Chimpanzee tool-use should show non-functional variation. The most likely ape to use flaked stone tools in the wild might be an old (especially toothless) chimpanzee in far western Africa where hammers and anvils are commonplace.

What can be said in summary about this chapter's third aim? It seems likely that ancestral hominoids made and used tools no less complicated than those used by living chimpanzees. This suggests that some sort of material culture (and the term is used advisedly, as discussed in the next chapter) long pre-dated the first lithic artefacts recognisable in the palaeo-archaeological record. As Foley (1987*a*) stressed, we will only be able to begin to infer further the nature of ancestral hominoid life by doing more palaeo-socio-ecological analyses of extinct hominoids. Among the living hominoids, the chimpanzee seems to present the best heuristic source of knowledge of this exercise, and this is pursued in Chapters 7 and 8.

4

Cultured chimpanzees?

Introduction

Can the concept of culture be applied validly to another species? This chapter reports a kind of grooming shown by wild chimpanzees which seems to be a truly social custom. The example serves to demonstrate the practical pitfalls and potentials of seeking to answer the above question. The goal is to test the utility for our closest living relations of a higher-order concept originally defined for human beings. Findings from studies of chimpanzees and of Japanese monkeys force us beyond the usual hazards presented by anthropomorphism in its various forms. At the same time, these findings show that if concepts such as culture can help in the understanding of the behaviour of other species, one must avoid simplistic and sloppy extrapolation.

Gombe and Kasoje compared

As should be clear after the first three chapters, two long-term field studies of wild chimpanzees have proceeded in parallel in western Tanzania, and most of the published knowledge of the natural behaviour of individual chimpanzees comes from these. Goodall's (1968, 1986) research group in the Gombe National Park has focussed on the Kasakela community of chimpanzees, whose membership has fluctuated from 38 to 60 (Goodall, 1986, p. 80). The project begun by the African Primate Expedition at Kasoje in the Mahale Mountains, initially under the direction of Itani and later of Nishida (1968, 1990), focussed first on K-group, then later on M-group (see Figure 4.1).

For many reasons, these two longitudinal studies are ideal candidates for comparative studies. First, both are of the eastern subspecies of chimpanzee. This should reduce the chances that any differences found

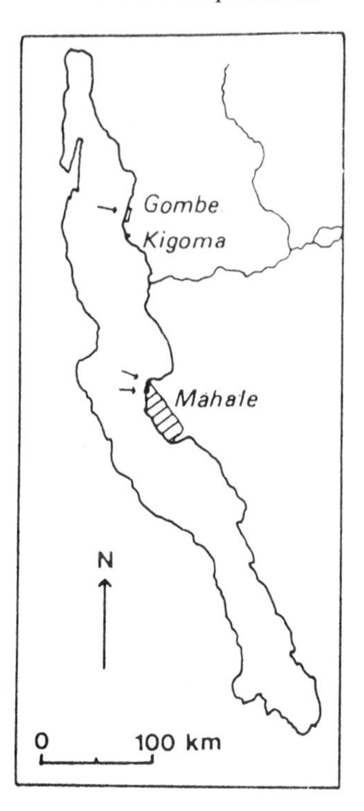

Figure 4.1. Locations of Gombe and Mahale Mountains on the eastern shore of Lake Tanganyika, Tanzania. (Courtesy of A. Collins.)

between them are genetic, as might be more likely if they were of different subspecies. Second, both are relic populations of what was once part of a continuous cline, since disrupted in modern times by deforestation (Kano, 1972). In straight-line distance the two study sites are only about 170 kilometres apart, and only 50 kilometres separate the southern limits of the population containing the Gombe chimpanzees from the northern limits of the population containing the Kasoje chimpanzees (Kano, 1972, p. 47). While the potential for interbreeding and interaction between them no longer exists, it is unlikely that the two populations have had time to differentiate markedly through genetic drift. Third, they occupy similar types of habitat: mixed forest and woodland on the rugged escarpment of the Great Rift which forms the eastern shore of Lake Tanganyika. Detailed ecological studies of the two sites (Collins & McGrew, 1988) show that much overlap exists in the types of vegetation, and this is reflected in much commonality in diet (Wrangham, 1975, 1977; Nishida & Uehara, 1983; but cf. Nishida *et al.*,

1983). Fourth, both studies have focussed on one or two communities or groups in which all members are individually recognisable and well known. Fifth, both populations have been studied using similar methods based on provisioning. This has enabled near-continuous human monitoring of all known chimpanzee activities at very close range. The differences between Gombe and Kasoje are far fewer than those between either of them and any other long-term study-site.

Case-study: Grooming

Grooming is the co-ordinated fine manipulation, sometimes linked with the use of lips or tongue, and close inspection of the body surface of the self or of another individual (see Figure 4.2). In many kinds of primates, including humans, grooming has at least two distinct but compatible functions: on one level it is hygienic, serving to remove ecto-parasites, extraneous matter and bodily products. On another level, social grooming is an intimate interaction between two friends. It may be unilateral (A grooms B but not the reverse), mutual (A and B groom one another

Figure 4.2. Typical social grooming by one adult male of another, Kasoje, Tanzania. (Courtesy of C. Tutin.)

simultaneously) or reciprocal (A grooms B, then B grooms A). The social function of primate grooming has long been recognised (Yerkes, 1933), but three points about chimpanzees' grooming need emphasising. First, chimpanzees are keen to groom, as shown by Falk (1958) in a simple but elegant experiment in which opportunity to groom was offered as a reward. Second, chimpanzees do not randomly distribute their social grooming among their associates. Instead, at least adult males show preferences for grooming partners from whomever is present (Simpson, 1973). Finally, normal chimpanzee grooming shows a species-typical form which is the same for all known wild populations. Variants on the standard form can be seen both in the wild (Goodall, 1973, p. 713) and in captivity (McGrew & Tutin, 1972), but these are idiosyncratic elaborations.

In January 1975, Caroline Tutin and I watched K-group's chimpanzees at Kasoje for 11 days. We totalled 33.5 hours of observation over 10 sessions ranging from 1.5 to 4.5 hours each. All data were taken at 5–25 metres away, using descriptive scan-sampling, in which one of us tape-recorded a running commentary and the other timed or photographed events. At that time, K-group was the best-habituated and longest-studied group of chimpanzees in the Mahale Mountains. It numbered 28 individuals of whom we saw 26. The mean party size per quarter-hour sample was 9 (N = 128, range = 1–18). Comparisons of behaviour made across individuals take account of this variable attendance, in terms of the number of quarter-hours in which each chimpanzee was present. During the observations the apes engaged in the full range of their normal activities, such as eating, sleeping, fighting and mating.

To our surprise, we witnessed a behavioural pattern which at that time had not been described for chimpanzees. This we called the *grooming-hand-clasp* (see Figure 4.3). It always occurred at the beginning of or during an otherwise normal bout of social grooming. Each of the participants simultaneously extended an arm overhead and then either or both grasped the other's wrist or hand. Meanwhile, the opposite hand was used to groom the other individual's underarm area revealed by the upraised limb. In doing so, the two chimpanzees sat facing one another on the ground in a symmetrical configuration. Either both raised their right arms and groomed with their left, or vice versa. The effect was striking.

Some aspects of the behaviour deserve further description (for statistical details see McGrew & Tutin, 1978):

(1) With one exception, participants engaged in mutual grooming in pairs. The exception occurred when one adult male joined another in grooming the underarm of an adult female.
(2) Only adults and adolescents performed the pattern. Nine of 17 individuals in these age-classes did so, while none of the 10 younger ones did.
(3) Performance of the behaviour was evenly distributed in the group. Individual frequencies were highly positively correlated with the length of the observation time, suggesting that it was a regular activity.
(4) Both sexes showed the behaviour to the same extent, but the pairs tended to be of mixed-sex rather than same-sex composition.
(5) Bouts lasted an average of 15 seconds. On three occasions participants changed arms and carried on in a second bout.
(6) No preference for right or left arm emerged.

How important is the grooming-hand-clasp in daily life? It occurred at an average rate of once every 2.4 hours (at least in this sample), which exceeds those for tool-use, predation, food-sharing, and almost all sexual and agonistic behaviour. Against this must be set the *total* absence of the grooming-hand-clasp among the chimpanzees of Gombe over thousands of hours of observation in 30 years. It is simply unknown there. The two nearest behavioural patterns known at Gombe are: (*a*) brief use of an upraised arm in grooming invitation – but only one individual does this at a time and the upraised arm is not touched by the other (Goodall, 1968, p. 264; Plooij, 1978*a*), and (*b*) more prolonged grasping of branches overhead while engaged in social grooming. More importantly, the chimpanzees of Kasoje show social grooming which is otherwise typical of that seen in other wild populations of the same (Reynolds & Reynolds, 1965) and different (Tutin *et al.*, 1983) geographical races of chimpanzees, as well as that exhibited in captive groups (Merrick, 1977).

Only one other report of this behavioural pattern has come from other field-studies of apes. In the Kibale forest Ghiglieri (1984, pp. 145–6) found that it occurred in 38% of non-maternal grooming sessions shown by the Kanyawara chimpanzees in his secondary study-area. Most intriguingly, it was entirely absent from the grooming repertoire of the neighbouring Ngogo chimpanzees who lived 10 kilometres to the south-east.

What are the origins of these differences between neighbouring com-

Figure 4.3. Grooming-hand-clasp by two adults, Kasoje, Tanzania: (*a*) incomplete; (*b*) fully extended.

munities of chimpanzees? Galef (1976, p. 77) stipulated three means by which such a difference might occur: (*a*) genetically transmitted propensities which are virtually independent of environmental influence affecting their expression in ontogeny; (*b*) similarly structured transactions between individuals and their environment in one community as opposed to different transactions in another community; (*c*) transmission of behavioural patterns through social learning from one individual to another, according to the norms of the community in which they live.

It is possible that the behavioural differences seen between Gombe and Kasoje and between Kanyawara and Ngogo reflect differences in genotype. However, there seems not to have been enough time for genetic drift to account for the contrast. Alternatively, it is possible that the chimpanzees at Kasoje and Kanyawara are responding individually to a selection pressure present in their environment but absent at Gombe and at Ngogo. Such differences in the bio-physical environment exist between the populations: Kasoje's chimpanzees eat blue duikers

Figure 4.3. *cont.*

but Gombe's cannot, for the simple reason that duikers are absent at Gombe. However, it is hard to see how differences in habitat could account for the presence or absence of the grooming-hand-clasp. It is conceivable that Kasoje's chimpanzees suffer from (say) a bothersome axillary parasite which requires more frequent attention, but there is no evidence to suggest this. (The absence of higher frequencies of self-grooming or scratching of the underarms makes it improbable.) This leaves by exclusion the third possibility that the grooming-hand-clasp of Kasoje's chimpanzees is some sort of social custom.

Abundant examples now exist of the social transmission of acquired behaviour in groups of non-human animals. These have been demonstrated in quantitative observations and experiments in both laboratory and field. (See Galef (1976) for a review of vertebrates but cf. Galef (1990) for caveats; Nishida (1987) for primates.) Such findings may be equally impressive even in song-birds, such as milk-bottle opening by tits (Hinde & Fisher, 1952), but these cases are excluded from this discussion as being only indirectly relevant to questions of hominoid evolution. Objections in the earlier literature to biassed anthropocentrism (Kroeber, 1928) or simple reliance on anecdotes (Hart & Panzer, 1925) need no longer apply. The question now becomes: *Do such social traditions in animals satisfy accepted anthropological criteria so that these may be termed cultural?*

Defining culture

To begin to answer this question required some definition of terms. Before undertaking this, one must dispose of the red herring of near-synonymity in terminology, which has sometimes been used to skirt round the problem. In one form this has meant using the word culture in quotation marks but without distinguishing definition (Kummer, 1971, p. 11). More confusing have been usages such as 'pre-culture' (Kawamura, 1972), or 'sub-culture' (Kawamura, 1959) or 'proto-culture' (Menzel *et al.*, 1972). In none of these examples do the authors justify the neologism, that is, they fail to provide distinguishing criteria for differentiating the patterns discussed from human patterns subsumed by the term culture (Kitahara-Frisch, 1977). The implication from the alternative term is that what other animals do is somehow less than or different from what humans do, but in an unspecified way. This is likely to be true, in some sense, but the coining of new terms is no substitute for explicit reasoning.

Somewhat ironically, the same point has been made forcefully by anthropologists claiming that culture is *by definition* a human prerogative. Kroeber & Kluckhohn's (1952) comprehensive review of concepts and definitions of culture made it clear that this was the traditional view. Most of the 168 definitions of culture compiled by them make use of terms which refer specifically to its human nature, and practically all of the remainder contain this feature by implication. From the content of the definitions, it looks as though most of the quoted authors never considered the possibility of non-human culture, presumably because no convincing evidence then existed of natural populations of other species showing behaviour resembling culture. Evidence for cultural capacities in captive chimpanzees *did* exist before recent field-studies (Hayes & Hayes, 1952), but it did not appear in anthropological journals. Those anthropologists (Kroeber, 1928; Hallowell, 1960) who were aware of laboratory studies like Köhler's dismissed them as insufficient, as did Köhler himself (1927, p. 266).

As knowledge of socially acquired behaviour from field-studies of other primates has become available, especially for Japanese macaques (see below) and chimpanzees (Goodall, 1964), many students of cultural anthropology have been slow to acknowledge it. Montagu (1968) ignored it and continued to postulate that culture is a species-specific human adaptation. Dobzhansky (1972, p. 422) cited the primatological findings but maintained that culture is uniquely human without further comment. Mann (1972, p. 382) admitted that some definitions of culture are broad enough to apply to other species. He then proposed another system, termed 'human culture', based on learned behaviour that modifies the environment and that is crucial to survival. Holloway (1969) scathingly dismissed the idea of non-human culture, asserting that the paramount difference in kind and not degree between human and other animals is that of the uniquely human imposition of arbitrary form upon the environment. Weiss (1973) discussed the problem of human-ness and culture at great length, and gave a comprehensive historical review of the issues. Except for a passing reference to a popular periodical, he ignored the evidence from field primatology, so his re-assertion that culture is uniquely human comes as no surprise. Moore (1974, p. 537) argued that the prevailing concept of culture in anthropology 'makes more sense as ideology than as empirical science', citing intra-disciplinary conflicts among schools of anthropological thought as the origin of debate on the nature of culture. (For a recent reiteration of these themes see Carrithers (1990), who explains the origins of 'human cultural variability' in terms of 'human sociality'. See also Bloch (1991).)

The overall impression is that until recently anthropologists either long ignored the evidence for non-human culture, erected *ad hominem* criteria which avoided taking the phenomenon seriously or, having considered the problem, felt it necessary to move the goal-posts. Even when testable elements have figured in the analysis, such as Mann's that human culture is crucial to survival while non-human culture is not, no real examination of the data followed. The treatment of the issue is reminiscent of the prolonged debate in the 1970s over the status of language, when human uniqueness on another front was challenged by signing apes (Fouts & Couch, 1976).

Defining culture as uniquely human raises several problems. First, doing so merely pushes the problem back a step and makes it one of defining human-ness. This may not be a topical issue when there is only one living species of human beings, but consider the hypothetical case of the discovery of a remnant population of earlier hominids. In seeking to discern whether or not these creatures were cultural, we could no longer rely on the supposedly clearcut human–ape division into cultural and non-cultural. Instead we would have to discard the prevailing generalisations based on an easy dichotomy as being simplistic. Discontinuities between living forms in a phyletic line sometimes lull us into forgetting about the many extinct intermediate forms which once made evolution look much more continuous (Hallowell, 1960).

Lest this argument be rejected as merely a hypothetical one, it is worth remembering that such problems have always loomed large for students of prehistory. They have been faced with a continuum of artefacts of increasing complexity and the need to decide at what point in the evolutionary past these can justifiably be called cultural (Holloway, 1969). Two classic examples of this were the then controversial Pliocene eoliths (Oakley, 1965, p. 5) and the osteodontokeratic culture of the southern African australopithecines (Dart, 1949). The problems involved in interpreting these objects are well known and formidable. One must decide whether such objects are artefacts or merely artefact-like natural objects. One must infer behaviour from its products. One must infer culture from incomplete evidence, given that most artefacts have probably perished. These handicaps have not so far stopped prehistorians from trying to re-create the existence of cultural capacities in ancestral forms (Dart, 1956), and they seem unlikely to do so in the future.

Some students of cultural origins may adopt the radical reductionist viewpoint that culture is a redundant concept, and not an explanatory one, as suggested by Foley (1991). This would entail a major paradigmatic shift. Steklis & Walter (1991, p. 161) in their attack on the reification of culture by 'culturologists' asserted that 'culture is no more than the behaviour of individuals plus the stored representations that consist in real neural assemblies resulting from the experiential inputs in socio-environmental settings'. Operationally (see below) this seems fair enough, provided access can be found to the representations and assemblies.

Another kind of reconstruction which seeks to elucidate the origins of culture in the evolutionary past comes from ecological anthropology (Cohen, 1968; Foley, 1984). It was proposed that during evolution culture somehow became the human mode of adaptation, so that culture can be defined as the human ecological niche (Wolpoff, 1971; Hardesty, 1972). It was argued that such a niche transcends other niches, opening up a realm of opportunities previously unavailable to other, more specialised species (Montagu, 1968; Swedlund, 1974). This brings its own problems: Swedlund (1974, p. 518) acknowledged that such reasoning may lead to an over-emphasis on culture as an explanatory device in prehistory because of its undeniably pervasive effect in later history. Hallowell (1960, p. 204) raised a similar point: It is hard to imagine all the aspects of human culture that presently exist as having arisen simultaneously in early hominid evolution. For instance, given a creature which made tools and showed incest avoidance but lacked speech and property rights, would we assign to it culture? It is just as hard to imagine a generalised selection pressure for some all-purpose trait called culture as it is for intelligence or for morality.

What emerges from this confusion is the need for an *operational* definition of culture, that is, one which stipulates properties which are empirically observable and measurable (Cafagna, 1960, p. 118). To be heuristic, such a definition should be comprehensive, designating both necessary and sufficient conditions. Unless these criteria are met, the increasing body of knowledge from field-studies of non-human primates, especially from apes, will only lead to more confusion, rather than giving clues to the key aspects of cultural evolution. Unfortunately, Cafagna (1960, p. 130) reckoned that no formulation of culture then available met the formal criteria for definition. Moore (1952; Anderson & Moore, 1962) amplified this issue in sometimes pointed detail, noting especially those definitions which 'are expressed in figurative language which makes it difficult to determine what their acceptance would entail' (1952, p. 253). The situation has hardly improved, so that, for example, Bloch (1991, p. 183) defines culture as 'that which needs to be known in order to operate reasonably effectively in a specific human environment'. There still seems to be no satisfactory and accepted definition of culture to use in attacking the question posed in the first sentence of this chapter.

Lest the preceding paragraphs seem to paint a totally bleak picture, it should be made clear that other non-primatologists have tried to tackle the problem. Harris (1964) developed a 'meta-taxonomy' of cultural parts which constitutes a comprehensive system. This scheme is logical, empirical and hierarchical, being ultimately based on the simplest possible behavioural unit (the 'actone'). Although Harris devoted little attention to the behaviour of other species (1964, pp. 173) he concluded that they had cultures of their own. Furthermore, he asserted that the differences between human and non-human cultures were matters of degree and not kind. However, Harris later (1979, p. 122–3) revised his views: Culture was said to exist in rudimentary form in many species, but human culture was said to be 'absolutely unique among all organisms'. In neither case did he offer guidelines on how to make specific comparisons across species.

So, what can be done? What remains, in the absence of an accepted working definition, is to take a first step towards it by being as painstaking as possible in abstracting those qualities of culture which are thought to be crucial. These can then be applied in evaluating data from a questionable, that is, non-human case. Happily, Kroeber (1928) did this over 60 years ago for chimpanzee behaviour. In discussing Köhler's (1927, p. 314) observation of 'dancing' by his captive chimpanzees, Kroeber (p. 331) stated that: 'If one ape devised or learnt a new dance step, or a particular new posture, or an attitude toward an object about which the dance revolved; and if these new acts were taken up by other chimpanzees, and became more or less standardised; especially if they survived beyond the influence of the inventor, were taken up by other communities, or passed on to generations after him – in that case we could legitimately feel that we were on solid ground of an ape culture.'

Assuming that Kroeber's dicta can be generalised to behaviour other than dancing (cf. Williams, 1980!), they amount to six criteria – *innovation, dissemination, standardisation, durability, diffusion, tradition* – that together form the beginnings of an operational definition of useful stringency (see Table 4.1). Conveniently these conditions form a logical chronological sequence. This is not the first attempt at proposing testable criteria of culture, but it is more rigorous and comprehensive than previous attempts (Kummer, 1971).

Table 4.1. *Conditions of criteria for recognising cultural acts in other species (from Kroeber, 1928; McGrew & Tutin, 1978)*

Innovation	New pattern is invented or modified
Dissemination	Pattern acquired by another from innovator
Standardisation	Form of pattern is consistent and stylised
Durability	Pattern performed outwith presence of demonstrator
Diffusion	Pattern spreads from one group to another
Tradition	Pattern persists from innovator's generation to next one
Non-subsistence	Pattern transcends subsistence
Naturalness	Pattern shown in absence of direct human influence

Japanese macaques

Before going further, it must be said that all six of the above conditions have been satisfied by Japanese macaques. Reports of studies by Japanese primatologists first appeared in English 35 years ago (Imanishi, 1957), and Western recognition of their implications for cultural anthropology soon followed (Frisch, 1959). The studies of what was termed 'acculturation' or 'sub-culture propagation' of sweet-potato washing, wheat-sluicing, candy-eating, etc., are well-known and will not be detailed here. (But note that all three involve *processing* as well as just *eating* new foods.) Several review articles on the subject have been published at various stages of progress (Kawamura, 1972; Itani & Nishimura, 1973; Nishida, 1987). Questions arise from these findings, however, which suggest that Kroeber's six conditions are not enough. Almost all of the documented examples from Japanese monkeys, including the best-known ones specified above, result from direct human intervention into the natural lives of the monkeys. All stem from their being provisioned over long periods with a variety of human foods: this in turn builds upon the monkeys' inclination to raid human crops. Further, even cases of non-feeding acts such as the hot-spring-bathing (Suzuki, 1965, p. 67) and sea-bathing (Galef, 1990) developed from intentional shaping of behaviour through provisioning. Green (1975, p. 309) even suggested that differences between troops in vocalisations (dialects?) may result from inadvertent conditioning of individuals during the provisioning procedure, rather than from social learning.

This is *not* to say that all apparently cultural behaviour shown by Japanese monkeys results either directly or indirectly from provisioning. Stephenson's (1973, p. 66) quantitative analysis of across-troop differences in courtship is a case for which it is hard to see such a connection. Moreover, there is a confounding of two variables: it is provisioning (variable A) which permits close-range observation (variable B). Negative results from unprovisioned troops may merely reflect poor or sporadic conditions of observation. However, the overall conclusion stands that traditional behavioural patterns of Japanese monkeys seem to depend heavily on provisioning (Galef, 1990). Thus, they do not yet represent a sufficient test of Kroeber's six conditions.

Additional conditions for culture

If the aim is to seek in non-human culture clues to processes which may have operated in human evolution, then one should minimise or better yet eliminate cases of artificial influence from humans. This is easier said than done. Examples of invention and acculturation from captive primates are fascinating and illustrative of the adaptive capacities of other species (Eaton, 1972; Menzel, 1972, 1973; McGrew *et al.*, 1975). These cases may tell us much about cultural processes in controlled conditions, but it is hard to see how snowball-making or use of ladders and pitons could illuminate behavioural patterns that are responses to natural selection pressures. Instead these examples may be ingenious responses by intelligent organisms frustrated by boredom in unstimulating environments that pose artificial challenges. (To give a human example, how we cope with jet-lag is a fascinating and important case of adaptation, but it is hard to see how it could tell us anything about our evolutionary past.)

More complex problems may arise with wild primate populations living in habitats significantly modified by human intervention. In addition to the problems of provisioning *per se* (Chapter 2; Frisch, 1959, p. 594; Reynolds, 1975; Wrangham, 1974), others emerge when humans introduce agriculture or forestry. Two examples from Kenyan baboons illustrate this: One population showed a direct response by devising novel diversionary tactics as part of crop-raiding (Maples, 1969; Maples *et al.*, 1976). Another population on a ranch cleared of large carnivores responded indirectly by enlarging their carnivorous propensities (Harding & Strum, 1976). (Note that these examples, as with so many others, concern changes in foraging.) Carried to its logical extreme, this line of reasoning founders, however. It is unlikely that any population of wild primates now exists in a state unaffected by human activity, and even the presence of a field-worker may be enough to alter the subjects' activities. A state totally unaffected by human influences would be unnatural anyway, given humankind's legitimate place in the biosphere for millennia as a gatherer-hunter. Only after the domestication of plants and animals did the human species start to make an unnatural impact on wild primates.

Given these corollaries, two further conditions can be added to the six abstracted from Kroeber: *Non-subsistence* is solitary or social behaviour which transcends subsistence activity, so that it is not concerned with the capture of energy or other nutrients. Such non-subsistence activities are unlikely to be correlated with the distribution of resources in the environment (Galef, 1976, p. 79). *Naturalness* is behaviour shown by other species living in conditions in which direct human interference is minimal, and indirect human influences do not exceed levels exerted by human gatherer-hunters.

If any wild population of other primates could be shown to exhibit behaviour which satisfied all eight of these conditions it would seem hard not to grant them the status of cultural beings.

Chimpanzees as culture-bearers?

To what extent do wild chimpanzees meet these criteria? Using the Gombe population and its tool-use (Goodall, 1964, 1968, 1973) as convenient examples, the first six seem reasonably clearly demonstrated:

Innovation The problem here is to assign criteria based on negative evidence, that is, a behaviour can only be recognised as new after a long-enough period in which its absence is notable. At Gombe, all of the obvious candidates for culture, e.g. termite-fishing and leaf-sponging, were under way when studies began. However, Goodall (1968, p. 197) also gave examples of nest-building techniques which arose and enjoyed short-lived popularity. Such fashions could not be linked to any recognisable environmental changes. She also reported the invention of sticks as levers used to prise open cement and metal boxes containing bananas; this persisted as long as the boxes were available (Goodall, 1968, p. 207). The lack of more conspicuous innovation over an observation period of over 30 years suggests that chimpanzee society is culturally conservative (but see Chapter 7).

Dissemination Recognisable transmission of apparently socially acquired behaviour among peers at Gombe is rare. Impressive circumstantial evidence exists, however, for the transmission of patterns from older to younger individuals, especially from mothers to offspring (Goodall, 1973; McGrew, 1977). Galef (1976, p. 87) pointed out that the presumed mechanisms of observational learning enabling this transmission have yet to be shown. Recent results from laboratory studies suggest that this is tricky: chimpanzees may learn to use tools by observation but without imitation of specific techniques (Tomasello *et al.*, 1987).

To recognise dissemination requires that the behavioural pattern be more than idiosyncratic; it needs to be a norm, that is, shown by some significant proportion of the population. The proportion may vary with the pattern: for example, only adult chimpanzees may have the strength to dispatch mammalian prey, whereas all but the youngest infants may be able to build a nest. It seems sensible to ask that the pattern be shown by more than one matriline, thus excluding patterns shown by one exceptional individual and her offspring (Goodall, 1973, p. 165). At Gombe, all chimpanzees above the age of 2 years show termite-fishing.

Standardisation The degree of stereotypy or, in the case of material culture, the 'imposition of arbitrary form upon the environment' (Holloway, 1969, p. 395) is empirically a matter of extent. In making tools to dip for driver ants, chimpanzees choose and then modify certain raw materials (McGrew, 1974). Both the product and its use differ significantly from those needed for another, functionally related task, that of termite-fishing (Goodall, 1973, p. 157).

Durability The performance of an acquired behavioural pattern outwith the presence of the demonstrator occurs with all of the common kinds of tool-use at Gombe. This is easily seen, as the fluid pattern of day-to-day social life means that all mature chimpanzees spend much time alone. By the time an individual stops travelling constantly with its mother in late childhood or early adolescence, it shows proficient tool-use when alone.

Diffusion The spread of behavioural patterns from one community to another has not been seen at Gombe, though migration between communities occurs. This could be a function of the rarity of cultural innovation, or of the lack of parallel, simultaneous study of two or more communities. All of the major kinds of tool-use existed in indistinguishable form in the Kasakela and Kahama communities. The two groups shared the same range before the smaller Kahama community shifted

Figure 4.4. Infant plays with mother's discarded tool while she fishes for termites, Gombe, Tanzania.

away in 1971, so no past diffusion need be posited to explain their similar behavioural patterns. However, at Kasoje the first stage of diffusion *has* been seen: two immigrants into M-group have been seen to use bark tools to fish for termites (Takahata, 1982). They moved residence from a termite-fishing group (K) to a previously non-termite-fishing group (M), thus setting the stage for M-group's members to learn from them.

Tradition Persistence from one generation to the next is present for all major acquired behavioural patterns at Gombe, that is, they persist in offspring of known origin after their parents have died. Strong circumstantial evidence exists for at least one case of persistence through three generations of the F family: Flo, Fifi and Freud (McGrew, unpublished data) (see Figure 4.4).

Non-subsistence The grooming-hand-clasp as practised by the Kasoje chimpanzees in contrast to their Gombe counterparts fulfils the condition of non-subsistence. Another example which contrasts the same two populations is the leaf-clipping display of Kasoje's chimpanzees

(Nishida, 1980*b*, 1987). Kasoje's males tear up leaves in courtship, as subtle tools for capturing the attention of females. (These same sort of leaves are also used in other contexts as food, napkins or sponges.) The sexual behaviour of Gombe's males has been studied extensively (McGinnis, 1979; Tutin, 1979), but no such leafy courtship has been seen there.

Naturalness Only one study of wild chimpanzees fully satisfies the condition of naturalness. Despite extended efforts at Kasakati (Izawa, 1970), Budongo (Reynolds & Reynolds, 1965; Sugiyama, 1973), Kibale (Ghiglieri, 1984) and Assirik (Tutin *et al.*, 1983), most studies of non-provisioned chimpanzees have not yet yielded consistent, close-range observations of behaviour. The exception is the study of Boesch & Boesch (1989, 1990) at Tai, where noisy nut-cracking enabled subjects to be found and habituated, so that the other conditions could be tested.

Goodall (1964) did see several kinds of tool-use by the Gombe chimpanzees such as termite-fishing, leaf-sponging and ant-dipping *before* she began provisioning, thus their performance cannot be due to direct human influence. However, the vital details of these activities only emerged after the use of heavy provisioning. Further, several authors have claimed that indirect human influences can be marked; for example, Kortlandt (1986) suggested that the palm-nut-cracking chimpanzees of Bossou may have learned the technique from watching local villagers cracking nuts in the forest.

In summary, no single population of chimpanzees yet shows a single behavioural pattern which satisfies all eight conditions of culture. However, all conditions (except perhaps diffusion) are readily met by some chimpanzees in some cases.

Culture denied?

In a thoughtful and provocative chapter, Tomasello (1990) tackled in detail the problem of whether or not chimpanzee tool-use is cultural. His conclusions are sceptical, if not totally negative. If he is right, then much of this book is suspect, so his views deserve close scrutiny. (Others have posed similar questions, e.g. Whiten, 1989; King, 1991; Lethmate, 1991.) On the basis of his work with captive chimpanzees (Tomasello *et al.*, 1987) he argued that there is no convincing experimental evidence of chimpanzee imitation. This, plus a similar lack of data on chimpanzee teaching, persuaded him that all of chimpanzee behaviour can be explained without invoking culture. This seems compelling, especially as he concluded that all potentially impressive cases of tool-use in the wild have been independently and *individually* invented by chimpanzees in the laboratory. Thus, candidates such as nut-cracking, leaf-sponging, termite-fishing and ant-dipping can be discounted. How solid is his case?

Failure to find true imitation under contrived and impoverished laboratory conditions is hardly surprising. Asking immature apes to learn from a brief exposure to a strange adult when they are evolved to learn from their mothers over years is expecting a lot (see Figure 4.5). Failure to find imitation in the field, when in principle it can never conclusively be proven under uncontrolled conditions, is illogical. A similar argument probably applies to failure to find teaching, though until this phenomenon is clearly defined in operational terms, its confirmation or otherwise will be messy. (Boesch & Boesch (1991) presented suggestive anecdotal evidence for teaching in nut-cracking, but there are alternative, non-teaching explanations for most of what they provided.) Thus the question of whether or not chimpanzees show imitation or teaching remains open.

But even if chimpanzees *never* imitated or taught, this does not require us to deny them culture. As Tomasello remarks, many behavioural differences across human populations persist without such mechanisms. Wynn (1992) gave several examples of skilled human operations that are acquired 'only' by observational learning. What (small) proportion of human knowledge is transmitted by intentional instruction? Interesting as the detailed mechanisms of information transfer may be to experimental psychologists (see Galef (1990) for a similar argument on the issue of 'tradition'), they play little part in discussions of culture in ethnology or prehistory.

Tomasello's (1990) claim that all notable tool-use practices of wild chimpanzees have been invented by chimpanzees individually in the laboratory is simply wrong. Sumita *et al.* (1985) *trained* zoo chimpanzees to crack domesticated nut species. Kitahara-Frisch & Norikoshi's (1982) single subject never learned to leaf-sponge, though she did dip for juice with some items of vegetation. No experiment has ever simulated any of the delicate probing tasks such as termite-fishing shown by wild chimpanzees, and all of the purported examples listed by Tomasello (e.g. Menzel *et al.*, 1970) are simple reaching tasks that are readily done by monkeys (Beck, 1973). However, even if captive chimpanzees did spontaneously show all of these patterns (and under conducive circum-

Figure 4.5. Interaction between infant and mother as she dips for driver ants, Gombe, Tanzania: (*a*) infant watches mother insert tool; (*b*) infant reaches to mother's arm during insertion; (*c*) infant reaches to tool during mother's pull-through; (*d*) infant mouths tool after pull-through. (Courtesy of J. Moore.)

Figure 4.5. *cont*.

stances this would hardly be surprising), it need not say anything about chimpanzee culture in the *wild*.

As Tomasello rightly points out, the challenge is to tackle empirically chimpanzee culture in nature, or at least in naturalistic, free-ranging settings. There are several promising ways to proceed. One is to perform field experiments in which crucial variables such as exposure to particular objects and others is manipulated. Hannah & McGrew's (1987) report of the spread of nut-cracking was *not* such an experiment, but it could easily have been. Another is to be more precise in operationalising key points to be looked for in field settings, as Russon & Galdikas (1992) have done for imitation in rehabilitated orang-utans. Finally, and probably most demandingly, new theoretical approaches, such as King's (1991) emphasis on social information transfer, rather than limited (and outmoded?) ideas such as imitation or stimulus enhancement, may prove useful.

Tomasello (1990) has rightly called to task those who use terms such as culture and imitation lightly, but their applicability to chimpanzee life has hardly been tested. Whether culture occurs without imitation, or imitation without culture, remains to be seen.

So are chimpanzees cultural or not? (Lethmate (1991) posed a similar question.) In commenting on McGrew & Tutin's (1978) original report, Washburn & Benedict (1979) thought not. For them, even if another species satisfied all eight conditions, it could not be granted cultural status unless it also had language. Language may be the most efficient means of natural communication yet devised and human beings may be the only naturally linguistic creatures (both debatable points), but why should language be essential to culture? (For a similar conclusion from a somewhat different argument, see Bloch (1991).) Both living pre-verbal humans and non-verbal non-humans readily learn new behavioural patterns without it (Bloch, 1991; King, 1991). Moreover, if language were a necessary and sufficient condition for the emergence of culture, then we would probably have to deny culture to evolutionarily pre-linguistic creatures ancestral to anatomically modern *Homo* (Davidson & Noble, 1989). This might mean excluding, for example, those responsible for the Shanidar flower burial (Leroi-Gourhan, 1975). This seems excessive. More likely, language emerged gradually in hominoid evolution, as did culture and most other traits that impress us, at different points in time and at different rates.

Socio-cultural anthropologists might be unwilling to attribute culture to chimpanzees until and unless the acts of the apes could be shown to have *meaning* for them (T. Ingold, personal communication). Assuming that meaningfulness can be demonstrated (as opposed to assumed) in human action, this sounds like a daunting challenge to the primatologist. But, before tackling it, the primatologist is entitled to ask what would be an acceptable and convincing demonstration, so far as the anthropologist is concerned. Having been told this, the chimpologist can then set about tackling the problem empirically, just as Byrne & Whiten (1988) did for deception. If no operational criteria are forthcoming, then the objection may be suspect as mischievous.

Finally consider the following thought experiment: Suppose that the grooming-hand-clasp had been described by someone like E.T. Hall (1959) for a *human* society in East Africa. Suppose that he presented ethnographic data exactly as here, contrasting the gestural repertoires of two neighbouring cultures. It would be accorded cultural status without questioning, and would dutifully be coded into the Human Relations Area File, to be used in future cross-cultural analyses. Where does this leave the chimpanzees of Kasoje and Kanyawara?

5
Chimpanzee sexes

Introduction

In 1974 I presented findings on faunivory, tool-use and food-sharing by chimpanzees to a Wenner-Gren Foundation symposium on the Great Apes. Some of the data were mine, but most were trawled from the treasure trove at Gombe begun by Goodall (1968, 1986). The paper was eventually published (McGrew, 1979), but in the intervening 5 years the picture changed notably, and it has changed even more so in the last 10 years.

The data reported in 1974 were the first to indicate differences between the sexes in an adaptive suite of hominoid subsistence activities. Several others sought to tackle the implications of these issues in the 1970s (Isaac, 1978*a*; Tanner & Zihlman, 1976; Zihlman, 1978). What follows is a synthesis and updating of their views from the usefully detached position of the armchair and of my views from the position of a chimpanzee field-worker. Case studies from Gombe will be used as convenient take-off points. The over-riding question is: *How would a proto-hominid population make the transition from sex differences in diet to sexual division of labour in subsistence*?

Sex is arguably the most important independent variable in evolutionary biology. It is one of life's few simple dichotomies, and leads to some equally stark consequences; consider the old saw that no organism is ever only *partly* pregnant. Given this, one might expect studies of behavioural sex differences in hominoids, that is, the phenotypic expression of behavioural traits ultimately linked to the two types of chromosome, to be straightforward. They are not.

Sex or gender? An aside

A major complication of studying sex differences is usually reserved for the behaviour (but not the structure) of our own species of primate: gender. The term is usually applied to acts that are thought to come from culture and not from nature. Thus gender differences are seen as proximately the result of socialisation into gender roles and (more extremely) as ultimately liberated from the phylogeny of sex differences in other primates. For example, in some Western countries, female infants are conventionally dressed in pink and male infants in blue for apparently arbitrary and purely cultural reasons. (However, *if* the function is to signal the sex of the infant, and *if* it turned out that some such custom of signalling occurs in all human societies, *or* at least in those in which the tell-tale genitals are covered by clothing, we might think again about the supposed solely cultural function.) So, is *gender* to *human*, as *sex* is to *animal*?

This simple analogy raises several interesting questions about human and non-human primates (see also Hrdy, 1981): Are all differences between women and men, or boys and girls, ones of gender? Are all differences in apes between females and males ones of sex? How can we establish or disconfirm the existence of gender in a species that yet tells us little about its mental representations? Can we ever eliminate sex (as opposed to gender) as an ultimate explanation for social differences between the human sexes? Can we determine when and to what extent gender differences superseded sex differences in the process of hominisation? Unfortunately, none of these is a sensible empirical question at present, but all may usefully be kept in mind considering what follows in this chapter.

Sex differences in diet: invertebrates

Goodall (1963, 1968) was the first to report that wild chimpanzees ate other animals. Among invertebrates, these represented six orders of insects: termites; ants; bees and wasps; flies; butterflies and moths; bugs. Only the first two types, the truly social insects, were eaten often enough to give systematic data. Gombe's chimpanzees also ate the products of insect labour, such as bees' honey and termites' earth. One species from each order of social insects is scrutinised below as a type of prey. Chimpanzees at Gombe make tools from vegetation in order to exploit these species in termite-fishing and in ant-dipping.

Case study: Termite-fishing

The Gombe chimpanzees' technique of 'fishing' for termites was described first by Goodall (1964, 1968), and is considered in detail in Chapter 7, so it is only briefly reviewed here (see Figure 5.1). The chimpanzee first opens a hole on the bare surface of the termites' earthen mound. The ape then inserts into the mound a long, thin probe made of plant material such as a blade of grass, strip of bark or segment of vine. Most of these simple probes have been modified by the chimpanzee by shortening, splitting, stripping, etc. Unseen by the predator, the termites inside the mound attack the intruding object by clamping onto it with their jaws. The chimpanzee then carefully withdraws the tool and uses her lips to nibble the insects from it, often one at a time. The sequence is repeated many times in leisurely fashion.

Are there sex differences in any of the basic parameters of termite-fishing, such as frequency, duration, periodicity or efficiency? McGrew (1979) reported results drawn from almost 7500 hours of observation over 19 months from July 1972 to January 1974 (see Table 5.1). These were the pooled results of 37 observers using one of Gombe's standardised data-sheets, the Travel-and-Group Chart. The data, which

Figure 5.1. Adult female fishes for termites, Gombe, Tanzania.

Table 5.1. *Sex differences in chimpanzees in fishing for termites at Gombe (from McGrew, 1979)*

	Females	Males	Total
Observation hours	3864	3597	7461
Observation sessions	835	608	1443
Mean duration of session (h)	4.6	5.9	5.2
Subjects	16	14	30
Termite-fishing (h)	166.25	50.75	217
Termite-fishing bouts	372	123	495
Mean bout-length (min)	26.8	24.8	26.3
Time spent fishing (%)	4.3	1.4	3.0
Sessions with fishing (%)	22	10	17

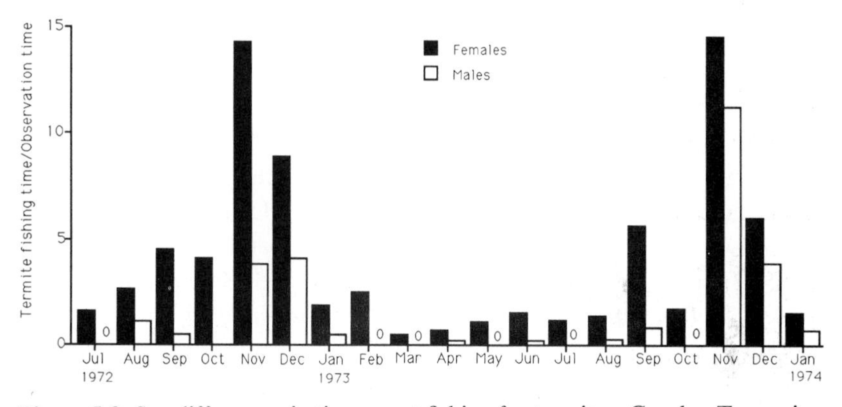

Figure 5.2. Sex differences in time spent fishing for termites, Gombe, Tanzania.

were based on focal-subject sampling, showed all eating bouts of a target chimpanzee to the nearest 5 minutes. The data comprised 1443 observation sessions averaging 5.2 hours' duration (range 0.5–13.25).

Thirty chimpanzees contributed to the data-set, ranging in age from 4.5 to 40+ years. They reflected the sex ratio of the population at the time, which was about 1 : 1. Overall the apes fished for termites for an average bout-length of 26.3 minutes (range 5–200).

The community as a whole averaged just under 3% of the observation time (which reflects waking hours) fishing for termites. This showed wide seasonal variation in monthly means from 0.25% to 13%. Overall, female termite-fishing occurred three times as often as male termite-fishing: 4.3% vs. 1.4%. In all 19 months the females' frequency exceeded the males' (see Figure 5.2). Looked at another way, termite-fishing occurred in 17% (range 1–57%) of observation sessions. In all 19 months females fished for termites in a higher proportion of sessions than did males.

Goodall (1968) suggested a difference between the chimpanzee sexes in termite-fishing by reporting that males were never seen to fish for more than 2 hours at a time while females often exceeded this. McGrew (1979) confirmed this: bouts of 60 minutes or more were done almost twice as frequently by females as by males. However, mean bout-lengths over all episodes were similar for the sexes (see Table 5.1). This means that female predominance in termite-fishing arises from fishing more often, particularly out of season, but not from longer bouts.

Other variables which might be responsible for the difference between the sexes, such as bias in methods or scheduling, were ruled out (McGrew, 1979). Thus, there appears to be a genuine sex (or gender) difference in termite-fishing by Gombe's chimpanzees.

Chimpanzees, tools and termites

Although chimpanzees using tools to obtain termites to eat has been recorded at many other sites in Africa (Kasoje: Uehara, 1982; Assirik: McGrew *et al.*, 1979*a*; Okorobikó: Jones & Sabater Pí, 1969; Belinga: McGrew & Rogers, 1983; Campo: Sugiyama, 1985; Bilenge: McGrew & Collins, 1985), none has yet produced behavioural data on differences between the sexes. At sites where such tool-use was common, the subjects were not habituated to human observation (Assirik), or at sites where subjects were habituated, the pattern was rare (Kasoje) or absent (Tai). Thus another approach is needed if comparisons across populations are to be made.

Faecal sampling provides an easy, quantitative method of assessing diet from its residues (Moreno-Black, 1978). It is particularly useful for detecting insectivory, as all insects have at least some chitinous body-parts that pass undigested through the chimpanzee's gut. The pooled faecal data from Gombe corroborate the behavioural data: Goodall (1968, p. 187) collected 194 faecal samples from 30 identified chimpanzees of juvenile age or older. Remains of food-types were noted nominally as present or absent. Over three times as many female samples contained termites as did males': 25% versus 8% (see Table 5.2).

Uehara (1986) has reported analyses of over 1500 faecal samples from two unit-groups of chimpanzees at Kasoje, but only five of these contained termites, so no analyses for possible sex differences are yet poss-

Table 5.2. *Sex differences in chimpanzees in eating vertebrate and invertebrate prey at Gombe, from analyses of faeces (from McGrew, 1979). Data are percentage of faecal samples containing prey*

	Females	Males	Total
Subjects	11	19	30
Faecal specimens[a]	81	113	194
Termites	25%	8%	15%
Weaver ants	32%	14%	22%
All invertebrates	60%	29%	42%
Vertebrates	1.2%	11.5%	7.2%

[a] Faecal data collected by Goodall (1968, p. 187), June 1964 – March 1965.

ible. Few faecal data have yet been reported from Tai (Boesch & Boesch, 1989). Thus to investigate further differences between the chimpanzee sexes in insectivory means turning to another type of prey: ants.

Case study: Ant-dipping

The tool-use technique of ant-dipping was described at Gombe in brief by Goodall (1963) and later in detail by McGrew (1974) (see Figure 5.3). In summary, the chimpanzee predator finds an underground nest of driver ants and digs into it by hand. The ape then makes a long, smooth wand of woody vegetation by modifying a branch. When the tool is inserted into the nest, the ants stream up it in attack. The chimpanzee quickly withdraws the tool and, while holding it in one hand, sweeps the length of the wand with the other in a loose grip ('pull-through'). The ants are momentarily collected in a jumbled mass on the sweeping hand and are directly popped into the mouth. The ape then chews frantically to avoid being bitten. In response to the massed, active defence of the ant prey, the chimpanzee predator shows various tactics of positioning and technique, such as perching on a bent-over sapling in order to remain elevated above the swarming mass of ants on the ground.

The same set of Travel-and-Group Charts used for termite-fishing provided data for comparison of the sexes (McGrew, 1979). Twenty-four chimpanzees were seen to dip for driver ants. None was younger than 4 years. (Younger offspring tended to avoid the painfully biting

Figure 5.3. Dipping for driver ants, Gombe, Tanzania: (*a*) two males insert tools into the ants' nest (Courtesy of D. Bygott); (*b*) female completes pull-through and ingests ants. (Courtesy of S. Halperin.)

Figure 5.4. Infant waits overhead and out of range while mother dips for driver ants. (Courtesy of C. Tutin.)

ants, even while their mothers dipped for them (McGrew, 1977). (See Figure 5.4.) They did 75 bouts of ant-dipping that totalled just under 16 hours. The overall bout-length averaged almost 15 minutes, and the overall time spent at the ant-dipping site averaged almost 19 minutes (see Table 5.3).

Table 5.3. *Sex differences in chimpanzees in dipping for driver ants at Gombe (from McGrew, 1979)*

	Females	Males	Total
Subjects	11	13	24
Duration at dipping (min)[a]	20.2 (3–86)	16.7 (5–59)	18.7 (3–86)
Duration of eating (min)[a]	15.5 (3–48)	13.6 (5–32)	14.9 (3–48)
No. of times present at dipping sessions	60	67	127
Percentage present that dipped for ants	75%	45%	59%

[a] Values are the mean with the range in brackets.

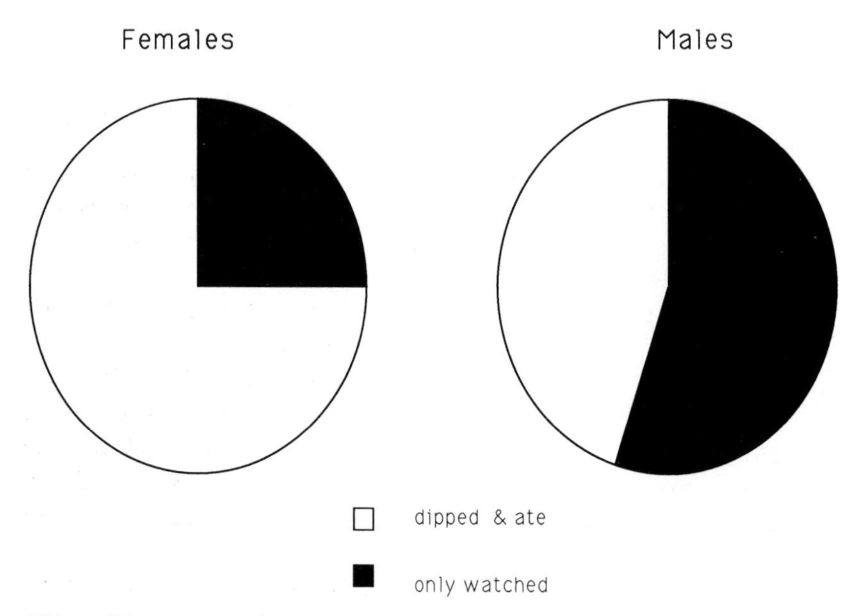

Figure 5.5. Sex differences in participation in dipping for driver ants, Gombe, Tanzania.

Given the smaller set of data, it is difficult to make as detailed an analysis of sex differences in ant-dipping as was done for termite-fishing. However, signs of sex differences emerged. Individuals aged over 4.5 years were seen 127 times at ants' nests when successful ant-dipping occurred. The proportions of those who actively dipped to those who only watched was very different though: three-quarters of the females dipped but less than half of the males (see Figure 5.5). As with termite-fishing, there was no sex difference in the mean length of a bout of eating ants, nor was there a significant difference in time spent at the dipping site. Only five samples of faeces contained driver ants, so no analysis of sex differences was done.

Chimpanzees and ants

Use of tools to dip for driver ants (see Figure 5.6) has been reported for several other populations of wild chimpanzees (Assirik: McGrew, 1983; Tai: Boesch & Boesch, 1990; Bossou: Sugiyama *et al.*, 1988). At Tai, females used tools successfully to dip for ants more often than did males; in contrast males used their hands (without tools) more often to scoop up pupae and larvae than did females. The smaller data-sets from Assirik and Bossou do not yet permit comparisons between the sexes.

However, by casting the net wider, more useful data on differences between the sexes can be added. Wild chimpanzees use a technique that is extractive but not instrumental to obtain weaver ants (Goodall, 1968; McGrew, 1983). They pluck the leafy nests and crush them, then leisurely peel away the leaves and consume the occupants of the natural container. At Gombe, 32% of females' faecal samples contained weaver ants but only 14% of males'.

At Kasoje, chimpanzees focus on two kinds of wood-boring ants: *Crematogaster* spp., which are extracted by breaking up dead twigs containing them, and *Camponotus* spp., which are fished with tools from tree-holes (Nishida, 1973, 1977; Nishida and Hiraiwa, 1982). Hiraiwa-Hasegawa (1989) reported significant differences between the sexes in time spent eating ants: females 6.5% versus males 2.9%. These sex differences occurred in adults, adolescents and juveniles, but not in infants.

Finally, at two sites overall differences between the sexes in insectivory have been investigated. At Gombe, 56% of females' faecal samples contained at least one type of insect, versus only 27% of males'. At Kasoje, both unit-groups showed a similar pattern: M-group, females 91% versus males 44%; K-group, females 86% versus males 64%.

Thus, the picture for insectivory, both specifically and generally, is clear: females in several widely separated populations of wild chimpanzees tend to specialise in a variety of prey species of insects. In no population of chimpanzees has the reverse been seen.

Figure 5.6. Driver ants (*Dorylus nigricans*); note the large soldier at the top.

Figure 5.7. Adult male eats head of red colobus monkey, Gombe, Tanzania.

Sex differences in diet: meat

Less commonly but more spectacularly, Goodall (1963, 1968) also reported carnivory by Gombe's chimpanzees. Teleki (1973) documented and illustrated this meat-eating in detail. Of the vertebrate classes, the prey were birds and mammals (see Figure 5.7). Several species of birds were taken, especially eggs and nestlings, but the data were too few for analysis by the sex of the consumer (Goodall, 1968, p. 189).

Case study: Mammals as prey

Chimpanzees at Gombe preyed on at least nine other species of mammals (see Table 5.4). If cannibalism is added, as well as the apparent occasional taking of human infants (Goodall, 1968, pp. 189–90), the number increases to 11. Other primates were overwhelmingly preferred: they were 259 of the 376 (69%) identified mammalian prey eaten over the period 1960–1981 (Goodall, 1986, p. 269; see Figure 5.8).

From the beginning of study at Gombe, males predominated in

Table 5.4. *Species of mammals eaten by chimpanzees at Gombe (from Goodall, 1986, pp. 268–9)*

Order	Scientific name	Common name	Major prey[a]
Primates	*Colobus badius*	Red colobus monkey	Yes
	Papio anubis	Olive baboon	Yes
	Cercopithecus mitis	Blue monkey	No
	C. ascanius	Red-tailed monkey	No
	Pan troglodytes	Chimpanzee	No
	Homo sapiens	Human	No
Artiodactyla	*Tragelaphus scripta*	Bushbuck	Yes
	Potamochoerus porcus	Bush pig	Yes
Rodentia	*Funisciurus* sp.	Striped tree squirrel	No
	Sp. indet.	Mouse or rat	No
Chiroptera	Sp. indet.	Bat	No

[a] Major prey are species comprising more than 5% of kills.

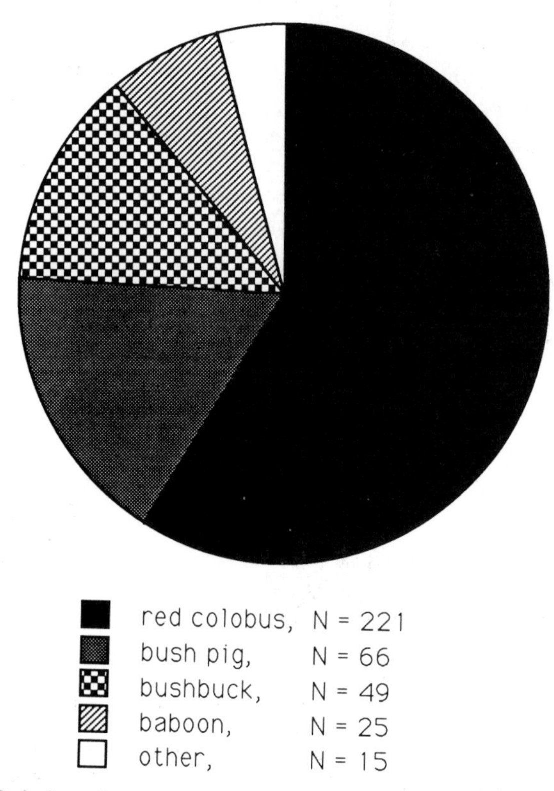

Figure 5.8. Relative frequency of type of mammalian prey eaten by chimpanzees, Gombe, Tanzania. $N = 376$ kills. (From Goodall (1986), p. 269.)

predation: in 1960–7, 28 kills of mammals were seen and all involved males (Goodall, 1968). In the 30 kills seen between March 1968 and March 1969, only adult males were seen to initiate predation and to pursue and capture prey (Teleki, 1973, p. 56). Adult males also divided the carcass, most often attended the feeding sessions after a kill, and ate most of the meat. Between 1972 and 1975 80 prey were taken, and Wrangham & van Zinnicq Bergmann Riss (1990) presented detailed analyses in terms of rates: males killed 36 times more often than did females over that period. The faecal data (see Table 5.2) confirm the behavioural data, showing that males eat meat almost 10 times more often than females.

Recently, Goodall (1986, p. 304ff) reported that Gombe's females get and eat more meat than was previously thought. She attributed the new finding to rectification of sampling bias but presented no evidence of this. The frequency of female participation was impressive: from 1974 to 1981 females obtained and then ate at least part of 44 prey. However, no rates of predation were given, so direct comparisons with males cannot be made. The only quantitative comparison across the sexes that is possible from Goodall's (1986, pp. 304, 310) data concerns the proportion of meat-eating sessions in which individuals ate 'some' meat. Males did so significantly more often than did females, and ate 'much' meat even more often than did females (see Figure 5.9).

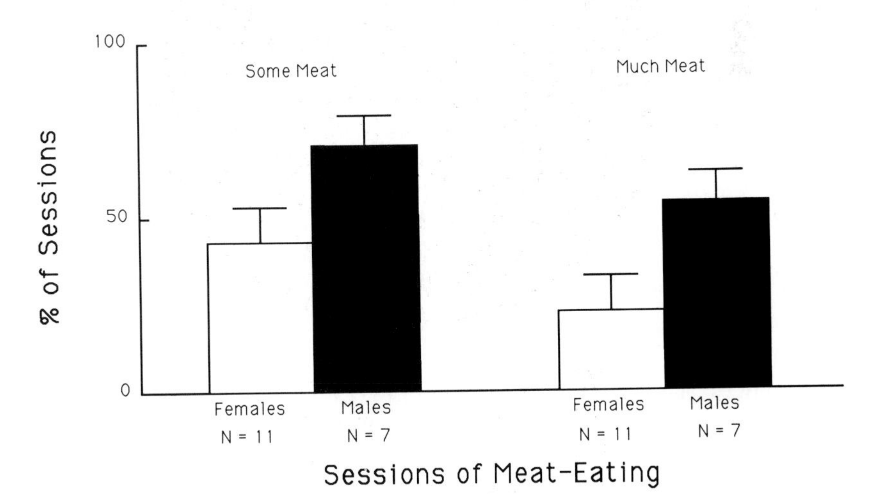

Figure 5.9. Sex differences in eating meat, as a proportion of the sessions in which some or much meat was eaten, Gombe, Tanzania. (Means and standard errors calculated from Goodall (1986) table 11.11, p. 304, and table 11.15, p. 310.)

Table 5.5. *Sex differences in chimpanzees in species of mammals preyed on at Gombe (from Goodall, 1986, table 11.2, p. 269; table 11.12, p. 305)*

	Red colobus	Bush pig	Bushbuck	Total
(*a*) Caught/eaten by females	24 (50%)	14 (29%)	10 (21%)	48 (100%)
(*b*) Caught/eaten by males	197 (68%)	52 (18%)	39 (14%)	288 (100%)
(*c*) Caught/eaten by chimpanzees	221	66	49	336

(*a*) = Columns 2–4 and rows 1–3; 1974–81.
(*b*) = (*c*) − (*a*).
(*c*) = Columns 3–5; 1960–81.

There is an apparent difference between the sexes in choice of prey, which may be related to method of capture. There are enough data (Goodall, 1986, pp. 269, 305) for comparison on three species of prey: red colobus monkey, bush pig and bushbuck. Red colobus are chased down in the upper canopy, despite sometimes offering fierce resistance. Bush pigs are taken on the ground where the young are cached in nests and sometimes defended by the adults. Bushbuck fawns are cached solitarily on the ground and only rarely defended by the mother. Males take a higher proportion of the more difficult-to-get colobus while females take a higher proportion of the easier-to-get ungulates (see Table 5.5).

Carnivory elsewhere

Since Goodall's (1963) first report of faunivory by wild chimpanzees, at least 10 other populations have been reported to eat at least 25 species of mammals (see review in Wrangham & van Zinnicq Bergmann Riss, 1990). For only two of these are there behavioural data available on sex differences: Tai (Boesch & Boesch, 1989) and K̇asoje (Nishida *et al.*, 1979; Takahata *et al.*, 1984). At Tai, females were less often present and active at predatory attempts, but when females did take part, their capture-rate was comparable to males'. At Kasoje, males were more often seen in possession of prey than were females; there was a significant preference by females to prey on the young of ungulates, probably by 'seizure', while males focussed on monkeys, usually by chasing.

Faecal data from Kasoje partly support this conclusion: in M-group males ate more meat than females; in K-group there was no difference (Uehara, 1986).

Thus, predation on mammals by chimpanzees is mostly a male activity.

Sex and faunivory

How are we to explain the female concentration on insects and the male concentration on mammals in the diet of chimpanzees? First, the behavioural patterns involved in getting the prey seem important. Male chimpanzees obtain meat by stalking, chasing, capturing, killing, dismembering, and distributing a prey animal. This often occurs socially, during the course of ranging widely with other males. In short, it is *hunting*. On the other hand, female chimpanzees typically obtain insects by prolonged, systematic, repetitive routines of object manipulation. Several individuals may forage together, but basically it is a solitary accumulation of a meal from many small units that are concentrated at a few permanent or predictable sources ('patches'). In short, it is *gathering*.

But *why* do chimpanzees show this sex difference? This is especially puzzling, given that both sexes are capable of exploiting both types of prey, at least in some contexts. It seems likely that selection pressures favouring sexual dimorphism, overlain on basic mammalian physiological adaptations, combine to favour such sex differences. First, the female co-option of the mammary glands as organs of nutrition for offspring meant that females became the primary parental investor in mammalian reproduction. Then, later selection pressures favouring rougher and tougher males produced greater body-size, physical strength and dental armament in males. Whether this came as a result of female selection for better defenders or from male–male competition for mates or both, the resulting difference in forms fits neatly with different specialisations in faunivory.

Males are probably better at hunting because their greater size and strength enable them to dispatch more and larger prey. At Gombe, these may be as large as an adult red colobus weighing over 8 kilograms. Larger canine teeth in males enable them to deal more effectively with the prey's anti-predator responses. (These should not be under-rated: see examples of surprising fierceness in red colobus in Busse (1977) and Boesch & Boesch (1989).) The same traits enable them to pirate prey from sympatric baboons (Morris & Goodall, 1977). Being unencumbered with dependent offspring, males are energetically freer to roam widely, thus increasing the chances of contact with wide-ranging prey, and to perform pursuit requiring speed and agility, often high in trees. Males are thus advantaged in what is a spasmodic, unpredictable and hurried activity that requires acute balance, sudden changes of direction and bursts of exertion. Having to carry offspring for most of their adult lives, either pre- or post-natally, hampers females on all of these fronts.

At Gombe, one unusual individual illustrates this contrast: Gigi is a sterile but otherwise normal adult female. Unconstrained by reproduction, she behaves more like a male than a female in many ways, such as ranging widely (Goodall, 1986, pp. 66–7). This has been especially true of hunting, as she has attended twice as many hunts and meat-eating sessions as any other female (Goodall, 1986, pp. 307, 310).

Females do better at harvesting a reliable, localised food resource such as social insects. For example, termite mounds may remain active for many years: mounds which were fished successfully by Gombe's chimpanzees in 1960 were still productive at least 21 years later. It seems likely that by the time a young Gombe chimpanzee begins to travel independently of its mother, it knows the location of scores of termite mounds in its range (see Collins & McGrew, 1985, 1987 for details of prey availability). Daughters go on travelling with their mothers for longer than do sons, so they may well be better informed about the locations of such resources than are their brothers. Keen termite-fishers at Gombe make use of circuits of termite mounds, going from one to another by the most direct routes (McGrew, unpublished data; cf. Boesch & Boesch, 1984*a*). Similar patterns probably obtain for ant-fishing trees used by Kasoje's chimpanzees (T. Nishida, personal communication). This strategy may be important for the optimising of energy budgeting (cf. Boesch & Boesch, 1983).

At the mound, termite-fishing is a sedentary and interruptible form of extractive foraging. The fishing needs little more than forelimb motion, and young infants may cling in the mother's ventral 'pocket' and sleep, suck or watch while she fishes. Older infants explore, play or try fishing, all in safety. The passive, limited defence of the prey presents no danger of pain or injury. In nursery parties, several mothers may fish at once, while play-groups of infants clamber about the mound, with no apparent effects on the fishing. If an infant is distressed or needs attention, the mother breaks off fishing to deal with it, then resumes her eating. A similar picture applies to fishing for *Camponotus* ants at Kasoje (Nishida

& Hiraiwa, 1982). *However, none of these aspects of infant-care is compatible with hunting.*

Dipping for driver ants presents a different set of problems. Though migratory, the prey bivouac for a few days, and most dipping sessions are re-use of ants' nests exploited at least once before. Chimpanzees return to known nests or find them by chance while otherwise foraging. At the nest, the dipper again taps a resource that is fixed during use. The chimpanzee adjusts its exposure to the prey by moving in and out of the range of the ant defenders, which is limited to a few metres. Younger infants usually cling tightly to the mother while she dips, moving about on her body to avoid being bitten by the ants (McGrew, 1974, 1977). Older infants stay outside the ants' defensive perimeter, watching from a safe distance or playing independently. The mother may retire periodically from the fray, only to return after a break to groom herself or to tend to her offspring. The technique is more demanding than termite-fishing, but it is still compatible with the competing requirements of child care.

Thus, social and (usually) terrestrial insects are an economical source of animal matter for a mother with near-constant child-rearing duties. Acquisition of the food-item depends on a repertoire of elegant techniques rather than on strength or speed. The behavioural patterns are energetically thrifty and self-paced. As females find termites throughout the year, even in the leanest of times, and as little energy is wasted in checking mounds, their focussing on such gathering makes calorific sense. Finally, by specialising in an alternative to meat, females may avoid direct and indirect competition with males for animal matter in the diet.

It is tempting to interpret this difference as a possible 'pre-adaptation' for the evolution of a system of sexual division of labour.

Nut-cracking

Differences between the sexes are not confined to faunivory, however, nor are they always neatly complementary as in female gathering and male hunting. Boesch & Boesch (1981, 1984*b*) have reported impressive sex differences in chimpanzees using hammers to crack open nuts. At Tai, *Coula edulis* nuts were common and relatively easy to open with wooden hammers; *Panda oleosa* nuts were rare and so hard that stone hammers and anvils were needed to open them. This meant that while *Coula* nuts were opened up in the trees or on the ground, *Panda* nut-cracking was confined to the ground. Efficiency of tool-use technique was recorded precisely in terms of number of hits needed to open a nut or of number of nuts opened per unit time.

For the simplest task of opening *Coula* nuts on the ground, there was no difference between the sexes. However, males preferred to crack the dried and more brittle *Coula* nuts at the end of the season, rather than the fresh and harder-to-crack nuts at the beginning of the season, while females were equally proficient at both times. The more demanding tasks of opening *Coula* nuts aloft or of opening *Panda* nuts at any time, females did both more efficiently and more often, at least by adulthood.

Unlike the contrasting demands of insect-gathering and mammal-hunting, nut-cracking seems to be equally amenable to both sexes. So, why is there such a marked sex difference? Boesch & Boesch (1984*b*) posed five hypotheses that might explain this: nutritional, social, attentional, motor and cognitive. None of these alone accounts for all of the findings, especially for the difference in efficiency of technique, nor are the five mutually exclusive. Some hypotheses address more ultimate than proximate levels of explanation, but all may boil down to selection pressures acting on reproduction: females who are more efficient at harvesting nutrients by 'offspring-friendly' techniques should enjoy higher reproductive success than those who are less efficient. (Brown (1970) advanced similar arguments to explain the subsistence activities of human primates.) The same selection pressures operate more weakly, if at all, on males.

If direct reproductive effort is crucial, then one might expect to see corresponding differences between the sexes in the distribution of food. That is, an organism which not only obtains and processes a resource more efficiently for its own sustenance but which also diverts some of the resource to provisioning its offspring will be doubly advantaged. Chimpanzees do all of these in food-sharing.

Food-sharing

Food-sharing has been reported in various species of non-human primates (see Feistner & McGrew, 1989, for a review), but only in chimpanzees is detailed knowledge available from both captivity and the wild (de Waal, 1989). Over 50 years ago, Nissen & Crawford (1936) experimentally investigated the sharing of food and of food tokens in captive pairs of juvenile chimpanzees. Teleki (1973) detailed the distribution of meat by Gombe's chimpanzees in 10 episodes of predation on mammalian prey (see also Suzuki, 1971). Teleki saw extensive transfer of parts of the carcass to many individuals by recovery, taking and requesting. Large groups of chimpanzees, including many who arrived after the kill, gathered around the male hunters and got meat from them. Attendance at the kill did not guarantee receipt of meat, nor was its distribution equitable or systematic, but some patterns emerged. Eighty per cent of sharing involved adults of both sexes getting meat from males. Female chimpanzees in oestrus were more successful in getting meat than were non-oestrous females. Transfers of meat among males did not strictly follow ranks in social dominance; instead success was highly positively correlated with age (Wrangham, 1975). Matrilineal kinship ties were also predictive of the patterning of meat distribution (although patrilineal ties remain to be studied). Both Goodall (1968) and Wrangham (1975) stressed that transfer of meat was not always peaceful and sometimes involved intense bursts of competition. Very similar patterns have been reported at Kasoje (Nishida, 1992).

Case-study: Banana-sharing

More frequently, wild chimpanzees share vegetable foods (Nishida, 1970; McGrew, 1975; Silk, 1978). These food transfers are a daily occurrence when chimpanzees eat rare or hard-to-process foods. The most detailed data come from Gombe's long-term records of non-agonistic sharing of bananas in the artificial feeding area. McGrew (1979) presented 21 months (January 1973–September 1974) of pooled data on 19 females and 18 males transferring possession of food (see Figure 5.10).

Sharing of bananas at Gombe occurred in all age and sex classes of chimpanzees. The vast majority of instances were between a mother and her infant, juvenile or adolescent offspring (see Figure 5.11). Food almost always passed from mother to child rather than vice versa. When the 37 chimpanzees involved were analytically paired in all 658 possible dyadic combinations, only 5% (N = 33) of these dyads had known ties of matrilineal kinship. Yet these related dyads accounted for 86% of the 457 recorded transfers of bananas. Of the remaining transfers between matrilineally unrelated individuals, the pattern was far from random: almost three-quarters of these were adult males giving bananas to adult females. Overall, distribution of bananas was not random by age: most recipients were infants (see Figure 5.12) and, overall, recipients were younger than donors in 88% (402 of 453) of cases, which is the reverse of the usual dominance relations.

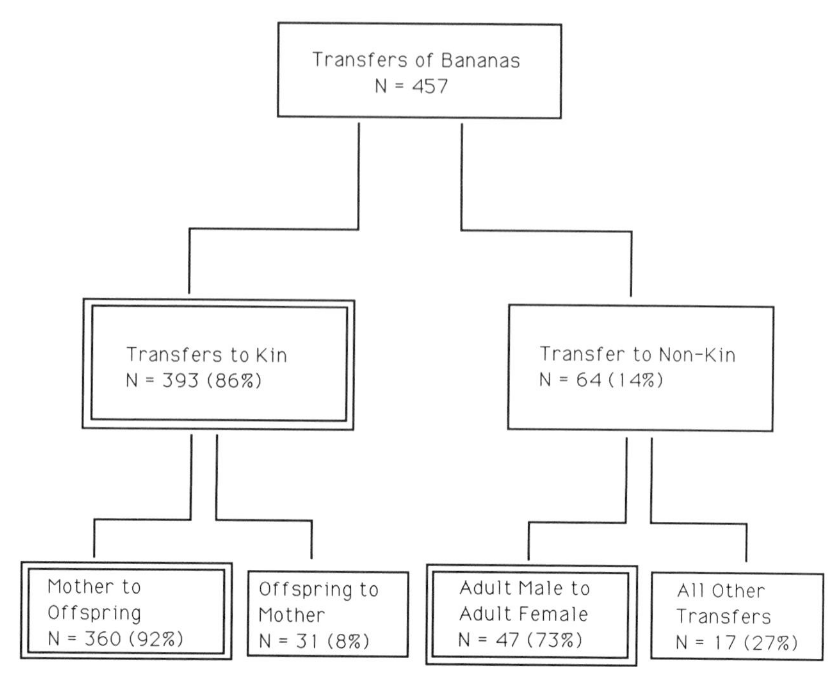

Figure 5.10. Patterns of transfer of bananas in food-sharing, Gombe, Tanzania. (In the transfers to kin, two cases of non-mother/offspring sharing have been omitted.)

Figure 5.11. Infant reaches to take bananas from her mother, Gombe, Tanzania.

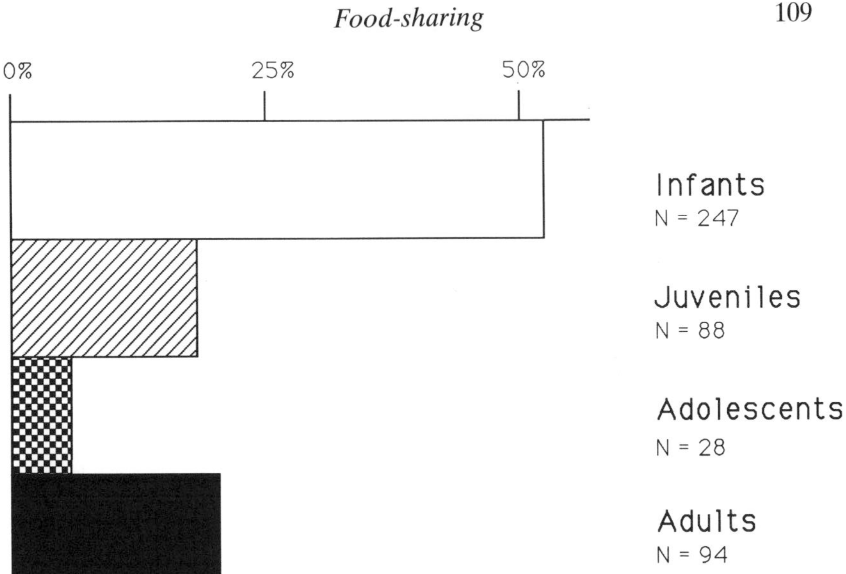

Figure 5.12. Age-classes of recipients of bananas in food-sharing, Gombe, Tanzania.

Figure 5.13. Adult male (right) shares bananas with adult female (left) with infant, Gombe, Tanzania.

Table 5.6. *Relationship between success of breeding chimpanzee females in getting meat and survivorship of offspring at Gombe (from Goodall, 1986, pp. 62, 310)*

Female	No. of living offspring	% of sessions to eat meat	
		Much	Some
Miff	3	34.0	63.0
Winkle	3	31.5	64.0
Melissa	4	31.0	52.0
Fifi	4	29.0	45.5
Athena	4	28.5	41.5
Mean	3.6	30.8	53.2
Nope	2	18.0	42.5
Pallas	0	14.0	25.5
Little Bee	2	7.0	26.5
Patti	1	6.0	12.5
Passion	2	0.0	28.0
Mean	1.4	9.0	27.0

Top vs. bottom 5 on Mann–Whitney U test, $n_1 = 5$, $n_2 = 5$, one-tailed: *Much*, $U = 0$, $p = 0.004$; *Some*, $U = 5$, $p = 0.075$.

Individual differences in the generosity of adult males emerged that were unrelated to ties of kinship (see Figure 5.13). All members of the community avidly consumed bananas, yet adult males rarely begged for them from anyone, and adult females almost never begged from one another. It seems likely that both sexes were playing out long-term reproductive strategies (cf. Tutin, 1979), but no systematic analyses of banana-sharing have been done. However, Goodall's data (1986, pp. 62, 310) showed a positive relationship between survival of offspring and sharing of meat at kills. The five most successful females at getting large amounts of meat had more surviving offspring than did the five least successful females, and the same contrast seemed to hold even with smaller amounts of meat (see Table 5.6). Similarly at Kasoje, the alpha male distributed meat mainly to females with whom he consorted, and to his mother (Nishida, 1992).

Observations on the sharing of naturally occurring fruits, found outside the artificial feeding area, confirmed these general patterns of sharing (Silk, 1978). For example, the hard-shelled, orange-sized fruit of *Strychnos* requires strength and technique to process for eating. No infant at Gombe was seen to accomplish this, but 2- to 5-year-olds cadged fragments from their mothers. The leathery pods of *Diplorhynchus condylocarpon* contained small amounts of edible seeds in sticky sap, and adult chimpanzees processed hundreds of these in prolonged sessions. The pod was neatly split in two with most of the contents ending up in one half; the mother often ate this half while passing on the other half to her infant. Infants did open their own pods, but they were messy and inefficient.

Other food-sharing

Many of the same non-random patterns of food-sharing by age and sex class have also been seen in Kasoje's chimpanzees (Nishida, 1970). Even details of technique, such as the mother dividing a fruit and giving a portion to her offspring ('halving behaviour'), were the same in both places. At Tai, Boesch & Boesch (1984*b*) noted that mothers always shared the results of their nut-cracking with their infant and juvenile offspring. At Gombe, termite-fishing mothers sometimes allow their infants to take termites from their tools, as well as extracted termites that had dropped to the ground (McGrew, unpublished data). Hannah & McGrew (1987) reported instances of sharing among unrelated adults in a released group of chimpanzees in Liberia: an individual with palm nuts gave them to another with a hammer, and both ate the proceeds of the processing (see Figure 5.14). In captivity, similar patterns of shar-

Figure 5.14. Adult male (right) reaches to take oil palm nut from the nut-cracker (left), another adult male, Bassa Islands, Liberia. (Courtesy of A. Hannah.)

ing, at least with youngsters, were observed in groups made up of wild-born chimpanzees and their progeny (Silk, 1979). Sharing of food seems to be basic to chimpanzee nature.

Other apes

How do chimpanzees compare with other apes in terms of differences between the sexes in faunivory, food-sharing and tool-use?

For orang-utans, all long-term field-studies have reported insects to be eaten commonly but vertebrates rarely (MacKinnon, 1974; Rodman, 1977; Rijksen, 1978; Galdikas & Teleki, 1981). None has reported statistical testing of possible sex differences in faunivory, and the untested results are mixed. Rodman's females spent over twice as much time eating insects as did the males. Galdikas found the reverse: males' exploitation rate of invertebrates was nine-fold greater than the females'. As with tool-use (see Chapter 3), food-sharing is virtually absent in orang-utans, being confined to tolerated scrounging by dependent offspring of left-overs from their mothers (Galdikas & Teleki, 1981, p. 245).

For gorillas, most data come from the highland population in Rwanda, where the only faunivory recorded is on invertebrates (Fossey & Harcourt, 1977; Watts, 1984). Quantitative data are few and as yet inconclusive, such as age differences in eating driver ants (Watts, 1989). Some lowland populations eat insects regularly, such as *Cubitermes* termites in Gabon (Tutin & Fernandez, 1983, 1985) or ponerine ants in Zaïre (Yamagiwa *et al.*, 1991), but no data on sex differences are yet available. The absence in gorillas of both tool-use and food-sharing other than tolerated scrounging resembles the situation in orang-utans.

For bonobos, data from the two main field-studies at Wamba (Kano & Mulavwa, 1984) and Lomako (Badrian & Malenky, 1984) agree closely: bonobos resemble chimpanzees in their range of faunivory from insects to ungulates. Unfortunately, quantitative data comparing the sexes are not yet available. Notably absent at both Lomako and Wamba is tool-use to obtain or to process either plant or animal foods. Food-sharing, however, is elaborate and habitual (Kuroda, 1984). Most data came from artificially provided pineapple and sugar cane, but sharing of natural plant (but not animal) foods was also seen. Differences between the two species of *Pan* outweighed the similarities: adult bonobos often shared food among themselves, food was often shared between non-kin,

Table 5.7. *Food getting and distribution techniques of great apes in nature, for plant, invertebrate and vertebrate prey*

	Plant			Invertebrate			Vertebrate		
	Eat	Tool	Share	Eat	Tool	Share	Eat	Tool	Share
Chimpanzee	++	++	++	++	++	+	++	+	++
Bonobo	++	−	++	++	−	−	++	−	+
Orang-utan	++	−−	+	++	−−	−−	+	−−	−−
Lowland gorilla	++	−	?	++	−	?	−−	−	?
Highland gorilla	++	−−	+	+	−−	−−	−−	−−	−−

++, well known in at least two populations; +, recorded at least once; −−, notably absent from at least populations studied long term; −, not (yet?) seen in limited or short-term studies; ?, not yet properly studied.

and only a minority of food-sharing seen was from mothers to their offspring.

Table 5.7 summarises current knowledge of the subsistence activities of the great apes in nature. There is both uniformity and variety across the five forms of great apes. All use extractive foraging (Parker & Gibson, 1979) to eat plant and insects, but only the chimpanzee is a significant predator on mammals. Similarly, only the two species of chimpanzee share food in ways other than the minimal tolerated scrounging of youngsters.

Origins of sexual division of labour

The constellation of food acquisition and distribution sketched above for chimpanzees is familiar, both in terms of the ethnography of human foraging peoples (Bird-David, 1990) and of speculation about protohominids (Isaac, 1978*b*). Comparison with the former will be done in Chapter 6, but here the question is: *How would a protohominid population make the evolutionary transition from sex differences in diet to sexual division of labour in subsistence*?

As a starting point in seeking to answer the question, it seems likely that individuals in those populations in which some members exploit different food resources from others will be more successful than those populations in which all members compete with one another for the same food resources. Regarding sex differences, this point was made long ago for non-primates (Selander, 1972), and it is now well known for primates (see review in Dunbar, 1988, p. 138). Such basic sex differences in diet might be thought of as incipient division of labour in that energy otherwise wasted in between-sex competition for food is therefore available for other activities.

For human primates, however, sexual division of labour means more than this: it means complementary activities that are consistently interdependent and that predictably benefit both sexes in a living group. Thus it is an example of *reciprocity* (Trivers, 1971). Such regular sharing of collected and processed food between individuals is likely to have been a key evolutionary development en route to modern human social organisation (Lovejoy, 1981).

Many authors have suggested that food-sharing arose in hominisation in connection with hunting (Etkin, 1954; Le Gros Clark, 1967; Tiger & Fox, 1971; Washburn & Lancaster, 1968). Because higher-quality animal foods are further up the food chain than are lower-quality plant foods, these are especially desirable commodities. At the same time, meat must be used quickly under tropical conditions before other organisms usurp it (Janzen, 1977). Food-sharing as part of sexual division of labour has sometimes been cited as the differentiating factor between apes and humans (Sahlins, 1959; Tiger & Fox, 1971). It may only be feasible for a male to gamble on hunting if he is bonded to a female who dependably produces surplusses from gathering which can buffer his failures to 'bring home the bacon'. Further, a male who provides animal protein for a pregnant or lactating female who is nurturing his genes should enjoy enhanced reproductive success (cf. Hrdy, 1981; Lovejoy, 1981; McGrew, 1981*a*). Thus, such reciprocal generosity need not involve sharing with kin to be selected for.

Contrary to some claims (Isaac, 1978*b*) wild chimpanzees show substantial food-sharing (see McGrew & Feistner, 1992, for a recent review). Both similarities and differences exist between chimpanzee and human food distribution of animal and plant foods. For example, like humans, chimpanzee meat distribution is not necessarily dictated by social dominance ranking. This suggests that later elaborations in distributive *roles* in groups of humans evolved from proto-hominid distinctions of status of greater complexity than 'Might makes right'. For example, sharing among male chimpanzees may help to sustain their cohesiveness, which is needed in other fronts such as the defence of the group's range against encroachment. Such sharing may increase the pay-offs from cooperative as opposed to solitary hunting or from more efficient defence of prey (see arguments for this carnivore-like analogy advanced by Schaller & Lowther, 1969). Evolution should favour those individuals capable of balancing competition and cooperation with peers by calculating the appropriate trade-offs of costs and benefits in particular contexts.

Chimpanzees may transport mammalian prey for more than short distances in various ways, apparently according to its size: quadrupedally by mouth or draped over the neck, tripedally with one hand supporting the prey slung over the shoulder. None of these methods looks efficient, and Hewes (1961) long ago pointed out the importance of bipedalism in the long-distance transport of food and other objects.

In contrast, almost all chimpanzee distribution of plant foods takes place at or near the source, except when the ape detaches a fruit-laden branch and retires a few metres to a more comfortable spot to eat it. The conspicuous exception occurs with the transport of a mouthful or a handful of nuts to an anvil for cracking with a hammer (Boesch & Boesch, 1983; Hannah & McGrew, 1987). Here the constraints of processing demand transport, as anvils cannot be taken to food.

Insect prey are not transported and are rarely shared. They come in the smallest natural packages of all, and it is only the larger-sized 'parcel' of the bees' comb which breaks this rule. It is both big enough to divide and rich enough in calories from honey or protein from larvae and pupae to be worth begging for and negotiating about. The closest thing to sharing of insects occurs when offspring eat from a mother's tool or pick up single prey overlooked or rejected by her.

What is noticeably lacking is what is arguably the single most important technological component of division of labour in subsistence: the *container*. Containers enable accumulation and transport of surplusses beyond individual needs, and these surplusses can then be shared (Ingold, 1986*d*). Chimpanzees spontaneously use containers in captivity (see Figure 10.2), so their absence in nature is not due to lack of intellectual appreciation of the principles involved. Nor is the absence of the container due to lack of raw materials; wild chimpanzees have access to leaves and suitable skins of mammals.

Origins of tool-use

Until recently, received wisdom in palaeo-anthropology linked the origins of tools in hominisation to the use of stone implements in the hunting and processing of large mammals by cooperating groups of males. For example: 'The first tools on earth were butchering tools' (Tiger & Fox, 1971, p. 121). '[Hunting was] . . . a master integrating pattern [which] . . . played the dominant role in transforming a bipedal ape into a tool-using and tool-making man who communicated by means of speech and expressed a complex culture' (Laughlin, 1968, p. 318). '[Hunting] . . . was presumably the principal factor that created the nuclear family' (Seward, 1968, p. 331).

Focussing on the hunting half of foraging led to the other half of the story, gathering, being under emphasised. Many human evolutionists ignored the importance of gathering or dismissed its products as 'casually collected foods' (Laughlin, 1968, p. 319). Coon (1971, p. 73) referred to the 'primacy of hunting' and stated categorically that it had more impact on social structure than did gathering. Lee (1968) and others criticised this viewpoint, not least because the conclusions were impressionistic, not empirical. Not long afterwards, the pendulum began to swing, and several workers began to focus on gathering as the key adaptation, notably Tanner and Zihlman (Tanner 1981, 1987; Tanner & Zihlman, 1976; Zihlman, 1978, 1981). By the early 1980s the balance was righted by an influential anthology, *Woman the Gatherer* (Dahlberg, 1981).

In the generalisations advanced in the speculative literature, many parallels exist between hunting by actual chimpanzees and by hypothesised protohominids: both are done mostly by males. Both concentrate on immature prey. Both parasitise other predators by piracy or scavenging. Both involve either solitary or social hunting. In social hunting both exchange information that helps to coordinate the actions of several hunters toward the common goal of bringing down the prey. Such communication is usually silent and inconspicuous, such as gazing and glancing, hair erection and stealthy locomotion. (Of course, coordinated action need not be cooperative. Individuals can act selfishly but simultaneously in ways that are hard to distinguish from collective action. See Boesch & Boesch (1989) for a detailed analysis.)

One major difference between chimpanzee and early hominid hunting is in the use of tools. Archaeological data from the Plio-Pleistocene onwards clearly show early hominids using tools in the processing of large, mammalian prey (see summaries in Toth & Schick, 1986; Potts, 1988). Although successful predation by chimpanzees on mammals has been seen almost 400 times at Gombe alone, only a handful of cases of tool-use have been seen (Goodall, 1986, p. 554ff; Plooij, 1978*b*). Some of the cases are hard to classify and seem no different from social or anti-predatory use of missiles, clubs and flails in display. At Tai, a young adult male broke off a branch to defend himself from a threatening group of red colobus. He first brandished it, then threw it at them (Boesch & Boesch, 1989).

More tool-use has been seen in the processing of mammalian prey, though only one technique is habitual: at Tai, in 26 of 28 kills, chimpanzees used small, modified sticks to pick out marrow from large bones broken open with their teeth (Boesch & Boesch, 1989). Other cases are anecdotal: Teleki (1973, pp. 144–5) described a male at Gombe using a wadge of leaves to clean out bits of brain left in the cranial cavity of a baboon being eaten. Boesch & Boesch (1989) reported a stick being used for the same purpose, and another case when a stick was used to clean the vertebral canal of a monkey's tail. S. Halperin (personal communication) watched an old male use leaves to catch his faeces as they were expelled. He fastidiously used the leaves as a 'plate' while he picked through the dung to extract undigested bits of meat which he then re-ingested.

Notably absent are cases of stone tools being used in butchery as hammers or slicers or scrapers. The closest to this comes again from Tai: three times chimpanzees smashed skulls of adult red colobus against a root or tree trunk. The motor patterns were the same as used otherwise to open hard-shelled fruits. Lack of true hammer-stone use to break bones by wild apes is not because they are incapable of the technique. Kitahara-Frisch *et al.* (1987) showed that chimpanzees in a zoo readily learned to use stone hammers and anvils to smash open the long bones of ungulates. The hollow shafts of pigs' femurs where the marrow would normally be were filled with a food-treat: chocolate. Whether and how captive chimpanzees, spontaneously or otherwise, would smash open bones to extract marrow remains to be seen. Such an experiment could be done easily in a number of captive colonies.

In contrast, the tools used by chimpanzees in obtaining social insects, as specialised in by females, are sophisticated. To take but one aspect of performance, Connolly (1974, p. 540) noted the greater skill involved in tool-use tasks when two hands are used in complemetary roles to attain a single goal. This occurs in ant-dipping: one hand holds the wand steady and upright in a power-grip while the other hand sweeps the length of the wand in a loose precision grip, catching up the ants in a jumble. When one hand is needed for suspension above the ants' nest, then a foot may substitute for the power-gripping hand, so that the process involves three limbs working complementarily. It is in this aspect of Table 5.7 that chimpanzees stand out so clearly: they are the only apes to use tools at all for natural subsistence tasks.

If the parallels between observed ape and hypothesised proto-hominid data are genuine, then the evolutionary origins of tool-use are more likely to have come from solitary, female gathering and not from social, male hunting. This idea dates back to Etkin's (1954) work, but has seen its most persistent exponent in Tanner (1981, 1987; but see also McGrew, 1979, 1981*a*).

A major problem in testing these ideas is that prehistory has come to be expressed almost exclusively in terms of stone tool morphology (see Chapter 9). Lithic artefacts were preserved better in the palaeo-archaeological record than were non-lithic ones such as bone, horn, shell, leather and, especially, plant parts. Yet the probability that non-lithic tools preceded lithic ones was long under-estimated: Oakley (1965) recognised that proto-hominids were likely to have used tools of bone and wood, but suggested that non-lithic materials came in only when stone tools were available to shape them.

Further, when functional assignments were made to early lithic tools, the presumption was that their use was in the processing of meat. For example, Leakey (1966) classed simple Oldowan hammer-stones as tools for smashing open long bones to get marrow. It is just as likely that they were used to smash open hard-shelled fruits or nuts or woody galls. The problem is the perennial one of negative evidence: it seems likely that the digging stick was a crucial component of the proto-hominid's tool-kit (Washburn, 1972), yet we are unlikely ever to find one in the palaeo-archaeological record because they have perished over millennia.

Chimpanzees at the best-studied field-sites, Gombe and Kasoje, do not use stone tools in subsistence; instead they use a varied tool-kit made of plant parts to obtain and to process both solid and liquid foods. For many of these tasks, such as the making of pliant probes, stone is not a suitable raw material (see Oswalt's, 1973, p. 12, discussion of 'flexibles' as a class of raw materials). However, Gombe's but not Kasoje's chimpanzees make use of anvils of stone or wood, against which they smash hard-shelled fruits (Nishida *et al.*, 1983). Many hammering blows may be struck before the rind of a *Strychnos* fruit cracks open.

Some far western chimpanzees, on the other hand, make use of stones as hammers (Boesch & Boesch, 1984*b*; Sugiyama & Koman, 1979). At Tai and Bossou they also use flexible tools of vegetation, but only minimally in ant-dipping and not in termite- or ant-fishing. The only population of wild chimpanzees to make skilful use *both* of flexible tools in fishing for social insects *and* of hammering against anvils of hard-shelled fruits is that at Assirik. (Whether or not these apes also use stones as hammers is unclear, as the evidence currently presented is only suggestive (see Bermejo *et al.*, 1989).) The population of chimpanzees at Assirik lives in one of the most marginal habitats in which chimpanzees survive in nature, which suggests that necessity may be the mother of invention (McGrew *et al.*, 1981; but cf. Kortlandt, 1983).

So what can now be said about the evolution of sexual division of labour based on what we know about sex (or gender) differences in great apes? Table 5.7 provides a useful framework. An omnivorous diet of plants and invertebrates was probably a basal feature of the ancestral hominoid, but there is no evidence from living apes that sex differences in diet were part of that adaptive package. Nor does it seem likely that vertebrates eaten at that stage were more than opportunistically encountered and of minimal nutritional importance, as is the case in all but chimpanzees today. Sharing of easily handled food-items was probably limited to tolerated scrounging. Neither for meat-eating nor for such simple food-sharing need there have been any sex differences, except as a by-product of maternal care for offspring. This basal pattern need not have been limited even to apes, as it fits some populations of savanna-living baboons equally well (Rhine *et al.*, 1986; Rhine & Westlund, 1978; but cf. Hausfater, 1975).

Chimpanzees and their proto-hominid counterparts show differences between the sexes which are more than by-products of features such as sexual or age dimorphism (cf. Post *et al.*, 1980, for baboons). Only in chimpanzees is there evidence of sexual *specialisation* in complementary animal foods (insects versus mammals) and in time-consuming or skilful processing of plant foods (nut-cracking). This is reflected in differences between the sexes in technical performance (tool-use); the correlation is unlikely to be merely coincidental. It is also evocative of what might be called a gender difference if it were seen in fellow human beings: the techniques of termite-fishing or nut-cracking are not favoured or constrained by any obvious features of sexual dimorphism, such as size or strength or a specific anatomical trait.

What chimpanzees lack is what may have been important in hominisation: *Tools for obtaining vertebrate prey and a means of collecting and transporting gathered food for exchange.* For the former, several candidates come to mind: club for dispatching well-defended prey (e.g.

porcupine); stick for digging up burrowing prey (e.g. pangolin); missile for disabling agile prey (e.g. monitor lizard); hammer for smashing well-armoured prey (e.g. tortoise); flail for downing roosting prey (e.g. fruit bat); hook for pulling down suspended prey (e.g. weaver bird's nest); etc. All of these techniques are within the capabilities of living chimpanzees and anecdotal evidence suggests that some are used. J. Hart (personal communication) working in the Ituri Forest of Zaïre reports finding sticks used to gouge out the contents of tortoise shells; the local pygmies attribute these to chimpanzees. Missing are tools that would bring down or at least slow down large prey, that is, wounding tools which can be used safely and effectively, such as a spear.

For food transport it is hard to see how a daily surplus can be achieved without a container in which to accumulate and to transport many small units (cf. Ingold, 1986*d*). All foods known to be gathered by chimpanzees are subject to these limitations; even cannon-ball-sized fruits like *Treculia africana* are awkward to carry for any distance.

Thus the transition from ancestral ape through chimpanzee-like proto-hominid to emergent hominid is readily imaginable, given the key evolutionary innovations of sexual division of labour and technology.

6

Chimpanzees and foragers

Cautionary note

Imagine a society characterised by the following: An extended family works together to collect and store a staple plant food upon which all depend collectively. Division of labour means that while some family members maintain the granary, others forage for animal prey which are brought back to a central place for sharing. All combine efforts to rear the family's offspring communally, with some members even deferring their reproduction in order to care for younger kin. If a parent dies, an outsider is recruited as a replacement mate, rather than incest being committed. Family life is a complex balance of cooperation and competition over many years.

This society is not imaginary but real. Moreover it is not a human, or even a primate, society but that of the acorn woodpecker (Stacey & Koenig, 1984). The plant food is acorns, and the animal prey is insects. These remarkable birds are mentioned here at the outset as a cautionary reminder that humans did not invent familial division of labour, nor are we necessarily its most impressive practitioners.

Why compare chimpanzees and hunter-gatherers?

Palaeo-anthropologists seeking to understand the evolutionary origins of human behaviour face a formidable obstacle. All of the players are long since dead and so are no longer behaving. Their stones and bones remain, and so provide a rich source of inference, but palaeo-psychologists must look elsewhere for acts and thoughts. The two main sources of data and ideas for these are modern foraging peoples and great apes. These supply the closest living approximations for calibrating the past process of hominisation. Sometime in the Miocene, an ancestral hominoid whose closest living analogue is an African pongid was set on a course that led to humanity. Sometime in the Pleistocene an early human being emerged whose closest living analogue is a tropical hunter-gatherer (or gatherer-hunter or forager). However crude, these two models of behaviour give us starting and ending points for reconstructing the process of human emergence. Each has been fruitful, but they seem not to have been seriously compared. (The closest thing to such a comparison is Peters & O'Brien's (1981) massive compilation of the plant foods of baboons, chimpanzees and humans across Africa. See also Peters & O'Brien (1982) and McGrew *et al.* (1982).)

Several reasons may explain this lack of comparison. Social and cultural anthropologists might think such a comparison to be a waste of time, believing the gap between human and non-human culture to be so wide as to be unbridgeable. Biological anthropologists might be content to make genetic, anatomical and physiological comparisons, but might baulk at the complexity and plasticity of behaviour. Archaeologists might agree in principle to compare the artefacts of ape and human, but might stumble in practice on the problem that most apes' tools are perishable while most surviving palaeo-artefacts are by definition the opposite. Primatologists, who come mostly from the natural sciences, might consider foraging *Homo sapiens* as fair game for comparison, but might be daunted by the paradigmatic leap required into the social sciences. However sensible, all of these obstacles are not good enough, as they are only *assumptions*, the validity of which is yet to be tested.

The basic point is this: *We will never know if such comparisons are useful unless we try them, however fraught they are with difficulties.*

Difficulties abound in both sets of ethnographic literature. Just defining hunting and gathering is problematical (e.g. Ingold, 1986*b*; Testart, 1988). Killing and sharing the meat of large land mammals may be undeniably hunting, just as collecting, processing and sharing the kernels of nuts may be undeniably gathering, but many aspects of the food quest are intermediate between the two poles. How are we to classify scavenging carcasses, poisoning fish, snaring birds, raiding bee hives, collecting turtle eggs, etc.? All of these subsistence activities are arguably a blend of gathering and hunting. Even plants as prey are not straightforward: some present fruits that 'beg' to be eaten, in that they are designed for fruit-eaters to disperse the seeds, while other plants produce deterrent toxins so potent that failure to process them properly may be fatal to the consumer. The closest thing to a safe generalisation that differentiates hunter-gatherers from other human societies may be the simple distinction between dependence on wild as opposed to domesticated organisms. (And of course even this is woolly when one considers the management of wild species or the exploitation of feral domesticated ones! See also Ingold (1986*b*).)

By the time that ethnographic data were collected, most hunter-gatherers were inextricably tied up with their cultivating or pastoralist neighbours (Headland & Reid, 1989; but cf. Solway & Lee, 1990) or were extinct. Instead of being independent foragers, they are now part of mutually dependent relationships that usually involve exchange of meat and labour for carbohydrates and goods (Bailey & Peacock, 1988). Further, in some (most?) cases, these symbioses are not recent but long-standing, as indicated (for example) by linguistic convergence (see Schrire, 1980, on the San). Even more sobering is the claim that some well-known hunter-gatherer societies are the creation of this mutuality. Tropical rain-forests, far from being primeval, may sustain foragers *only* when they can co-exist with agriculturalists (Bailey *et al.*, 1989). All of this means that simplistic views of hunter-gatherers as 'frozen in time', or 'living fossils', or 'windows on the past' are misguided (Lewin, 1988).

Living apes are at least easy enough to define and recognise, but their ethnography also presents problems. All field-studies of apes have focussed on populations beset by post-industrial humans. Forest clearance and hunting with firearms mean that few apes still live undisturbed life-styles in the environments in which they evolved. They are extinct over large portions of their recent ranges: for example, no orang-utans survive in mainland Asia. Further, most *behavioural* data come from subjects specifically tamed by scientists for study, as noted in Chapter 2.

Ideal versus actual comparisons

The best comparison would be of the closest living relations in the closest approximation to the environment of hominisation. More specifically, the ideal study would satisfy six criteria: sympatry, pristinity, simultaneity, methodological identity, longevity and comprehensiveness. That is, chimpanzees and hunter-gatherers would be studied in the same intact, tropical African ecosystem at the same time using the most similar methods over several annual cycles by integrated teams of research workers.

Unfortunately, no chimpanzee and hunter-gatherer populations have yet been *studied* in the same place. Few even *live* sympatrically, the only

Table 6.1. *Mean annual rainfall (to nearest 100 mm) for populations of hunter-gatherers and chimpanzees in Africa*

Population	Country	Rainfall (mm)	Sources
(a) Hunter-gatherers			
G/wi San	Botswana	400	Silberbauer (1972)
!Kung San	Botswana	500	Lee (1979)
Hadza	Tanzania	600	Terashima (1980)
Bambote	Zaïre	1100	Terashima (1980)
Ndorobo	Kenya	1400	Huntingford (1955)
Mbuti	Zaïre	1800	Ichikawa (1983)
Aka	Central African Republic	1800	Bahuchet (1978)
Efe	Zaïre	1900	Bailey & Peacock (1988)
(b) Chimpanzees			
Ishasha	Zaïre	800	Sept (1992)
Assirik	Senegal	900	McGrew *et al.* (1981)
Kasakati	Tanzania	(est.) 1000	Izawa & Itani (1966), Suzuki (1969)
Budongo	Uganda	1500	Eggeling (1947)
Lopé	Gabon	1600	Williamson (1988), C. Tutin & M. Fernandez (unpublished data)
Gombe	Tanzania	1800	J. Moore (personal communication)
Kasoje	Tanzania	1800	Takasaki *et al.* (1990)
Tai	Ivory Coast	1800	Boesch & Boesch (1989)
Okorobikó	Equatorial Guinea	2100	Jones & Sabater Pí (1971), Tullot (1951)
Bossou	Guinea	3000	Sugiyama & Koman (1987)
(c) Early hominids			
Olduvai	Tanzania (2.2 million years ago)	800	Cerling & Hay (1986)

eligible humans being the pygmies of equatorial forests (Baka in Cameroon: M. Harrison, personal communication; Mbuti in Zaïre: Ichikawa, 1983) and the Bambote of woodland Zaïre (Terashima, 1980). Foley (1982) has pointed out a notable gap in the distribution of extant African hunter-gatherers: they tend to live on arid savannas where mean annual rainfall is below about 500 millimetres (various San) or in forests where rainfall is above about 1500 millimetres (various pygmies). Most chimpanzees live in woodlands or forests where annual rainfall totals about 1500–2000 millimetres (McGrew *et al.*, 1981) (see

Table 6.1). Thus 'classic', open-country peoples such as the San or Hadza meet no apes, nor vice versa.

Since apes and hunter-gatherers now co-exist in disturbed habitats, they must compete for resources not just with one another (which would be instructive for evolutionary reconstruction) but also with agricultural, industrial and exotic human cultures. These forces are often repressive or destructive, so that survival may have to take priority, and both apes and foragers seem to be losing their respective battles. Both are adaptable, so that chimpanzees now raid crops and foragers now herd cattle, but this can hardly tell us about what took place in the Miocene.

Though studies of hunter-gatherers and apes are displaced in space, they have been roughly synchronous. The 1960s saw the explosion of interest in quantitative field-studies of behaviour, as exemplified by descriptions such as those by Lee & DeVore (1968) and Goodall (1968). This interest continued through the 1970s and 1980s (Dahlberg, 1981; Hamburg & McCown, 1979), though both may well have peaked, as funds have dried up, subjects have dwindled, and politics have limited access. Studies of both human and non-human foragers have benefited from the theoretical inputs of behavioural ecology and socio-ecology (Foley, 1984; Standen & Foley, 1989).

Methodologically, attempts at behavioural comparison are usually frustrating. Traditional, descriptive ethnographic accounts of hunter-gatherers tend to be qualitative and typological. More recent empirical efforts often supply more quantitative data on the results of behaviour (e.g. yields of prey) or on general daily activities (e.g. duration of hunts), but rarely reveal much about the acts of individuals. Few ethnological studies of African foragers meet modern ethological standards of definition, validity and reliability (Martin & Bateson, 1986), although there are exceptions (Bailey & Peacock, 1988). Surprisingly, the data from behavioural primatology are generally more rigorous, maybe because students of apes do not have the luxury of interviewing informants as well as watching them (cf. Wiessner, 1981). Such methodological mismatches may lead to confusion in comparisons – for example the idea that because fellow human beings can report their thoughts directly (truthfully or not) they act intentionally, whereas non-humans who do not disclose the contents of their minds must behave unintentionally (Ingold, 1986*a*).

Apes and foragers, even in equatorial rain-forests, live in seasonal environments. Thus, studies of less than an annual cycle are necessarily incomplete, and may be misleading, especially if there are ecological 'bottlenecks' through which organisms must yearly pass (Speth, 1987; Wiens, 1977). Further, tropical forests may show marked variability *between* annual cycles. Finally, environmental effects may be delayed, so that failure of this year's nut crop may result from last year's drought. All of these factors argue for long-term research, and few studies of African foragers qualify. Lee's (1968) much-cited quantitative input–output analysis of !Kung San subsistence was based on only a 3-week period (Wilmsen, 1982). There seems to be *no* study of African foragers which has lasted for 24 consecutive months or more. For chimpanzees, at least five field studies (Gombe, Lopé, Mahale, Tai, Assirik) have exceeded this duration, as described in Chapter 2.

Data on subsistence technology are only meaningful in terms of material culture in general, which in turn makes sense only in terms of culture on the widest scale, which is embedded in biotic and physical processes. Thus to understand hunting, we must know about spears, base-camps, rules of meat distribution and activity patterns of prey. This requires comprehensive, multi-disciplinary research teams, but these have been conspicuously absent, except in the Kalahari !Kung San Project in Botswana (Lee & DeVore, 1976) and the Ituri Project in Zaïre (Bailey & Peacock, 1988) that studied everything from intestinal parasites to dream interpretation. More notable are the gaps: after 30 years of research in the tiny Gombe National Park in Tanzania, a systematic census of the chimpanzee population has yet to be done (Goodall, 1986).

So, what comparisons can be made? A few reasonably comprehensive data-sets on chimpanzees such as those from Gombe and Kasoje can be compared with sparser data on African tropical foragers, especially pygmies and San. None is sympatric or pristine, and few methods are directly comparable or of extensive duration, but most were done at about the same time. The next section gives a very specific comparison between one group of chimpanzees and one group of San with regard to one aspect of environment: climate. This is followed by a case study in material culture, comparing food-getting tools in a matched pair of chimpanzee and human foraging societies. Finally, the chapter concludes with an overview of apparent similarities and differences between chimpanzees and African hunter-gatherers, in terms of questions that need further attention in order to bridge the gaps.

Hot, dry and open habitats: humans and apes compared

A persisting generalisation in textbooks on human evolution is that ancestral apes stayed in the forests while proto-hominids ventured out onto the savannas. Such a stark contrast, ultimately based on climatic differences, is usually central to explanations for the human–ape split in the Miocene. To test this ecological dichotomy one can look at exceptions, that is, at humans in forests and apes on savannas.

There are plenty of human foragers in tropical primary forests, but their history, much less prehistory, is little known. As stated above, it seems increasingly likely that hunter-gatherer occupation of forested niches is recent and dependent on neighbouring agriculturalists (Bailey *et al.*, 1989). More crucial is the presence of apes on savannas, of which there are at least three candidate populations, all chimpanzees: Bafing in Mali (Moore, 1985), Ugalla in Tanzania (Itani, 1979) and Assirik in Senegal (McGrew *et al.*, 1981). Only those at Assirik have yet been studied in the long term, and they can be compared with the best-known hunter-gatherers of the African savannas, the !Kung San (Lee & DeVore, 1976; Lee, 1979). In both cases there are two sorts of climatological data: limited data collected on site by the researchers during their studies, and longer series of data from nearby official meteorological stations (see Table 6.2).

Lee (1979) reported ecological data on the Dobe site in Botswana where the !Kung San were studied. The 3 years of annual rainfall data show great variation: 239, 597 and 378 millimetres, median = 378. This is drier than longer series from weather stations at Ghanzi, 250 kilometres south (Meteorological Office, 1975) and Maun, 300 kilometres east (Lee, 1979, p. 113), which average closer to 500 millimetres. At Assirik, 4 years of annual rainfall showed similar variation across years: 891, 824, 1224 and 879 millimetres, median = 885 (see Figure 6.1). This is similar to a longer series from a nearby weather station at Tambacounda, 140 kilometres to the north-west (Griffiths, 1972). Even given the high inter-annual variation, it is obvious that Dobe is drier than Assirik.

However, *distribution* of rainfall over the year reveals greater similarities. Here, calendar months are classed as wet or dry, with the criterion for dryness being a monthly mean of less than 1/24 of the mean annual total, that is, less than half of what would be expected by chance if rainfall were distributed evenly over the annual cycle. By these

Table 6.2. *Comparison of climatological data for savanna-living chimpanzees and hunter-gatherers (N = years of data)*

	Annual rainfall (mm)	Dry months	Totally dry months
!Kung San at Dobe, Botswana			
Dobe (N=3)[a] (19°35′S, 21°02′E)	378	6	5
Ghanzi (N=24)[b] (21°30′S, 21°45′E)	467	5	4
Maun (N=46)[b] (20°00′S, 23°26′E)	460	?	?
Chimpanzees at Assirik, Senegal			
Assirik (N=4)[a] (12°50′N, 12°45′W)	885	6.5	5.5
Tambacounda (N=35)[b] (13°46′N, 13°38′W)	872	7	6

[a] Medians for short series.
[b] Means for long series.

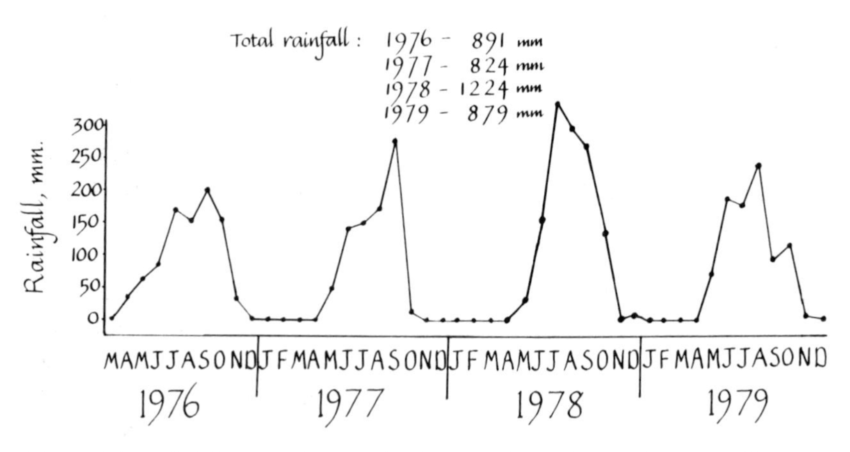

Figure 6.1. Rainfall month by month, Assirik, Senegal. N = 4 years. (Courtesy of P. Baldwin.)

standards, both the shorter and longer series of data show Assirik to be, if anything, drier than Dobe (see Table 6.2).

A more stringent criterion is number of *totally* dry months in the annual cycle, that is, months in which normally no rain falls. These are defined here as months in which the monthly mean is 1/10 of what one

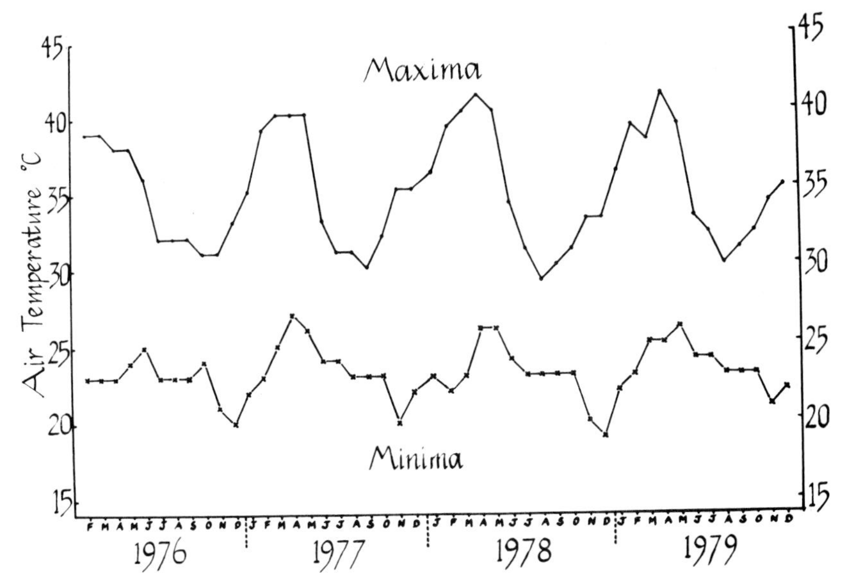

Figure 6.2. Temperature, mean monthly maxima and minima, Assirik, Senegal. N = 4 years. (Courtesy of P. Baldwin.)

would expect if rainfall were distributed evenly over the year, or only 1/120 or 0.8% of the mean annual total. Again, Table 6.2 shows Assirik to be drier, if anything, than Dobe.

Lee (1979, p. 106) also presented data on cold and heat stress experienced by the !Kung San at Dobe. For the cold, no quantitative comparison is needed, as the lowest temperature recorded at Assirik was 16 °C (McGrew *et al.*, 1981; see Figure 6.2) while in the Kalahari overnight low temperatures occasionally dropped below freezing. Comparison of heat stress yields surprising results. Lee (1979) counted as stressful those days in which the maximum air temperature reached 33 °C or more. Over 17 months of data, the !Kung were so stressed on 42% of days. At Assirik the comparable figure over 47 months was 80% of days (see Figure 6.3a). In the Kalahari, the !Kung San experienced heat stress on at least half of the days in 7 of 17 months of data; at Assirik the comparable figure for the chimpanzees was 31 of 47 months (see Figure 6.3b). In summary, the chimpanzees at Assirik suffer higher air temperatures than do the !Kung San at Dobe, but the latter must cope with lower temperatures.

The biggest difference between the two environments may be in the number of permanent sources of surface water for drinking. At Assirik

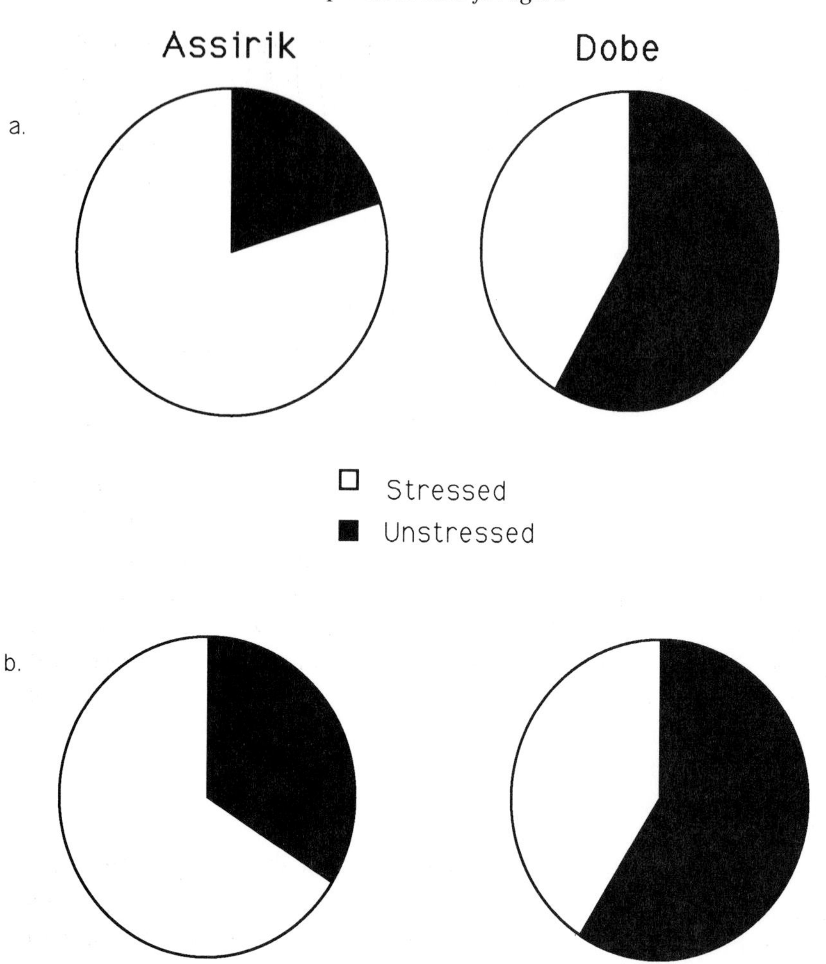

Figure 6.3. Comparison of heat stress at Assirik, Senegal, and Dobe, Botswana: (*a*) percentage of days with a daily maximum temperature exceeding 33°C; (*b*) number of months in which at least half of the days had a maximum exceeding 33°C.

there were six such sites within the 50 square kilometre core area of study. At Dobe there were only nine permanent water-sources within 11 000 square kilometres (Yellen & Lee, 1976). In both cases, availability of drinking water at the height of the dry season is likely to be the single most important environmental constraint.

Overall, the !Kung may occupy a harsher environment than that known for any population of chimpanzees. However, the climatological differences are hardly conclusive in terms of adaptation to savanna living. The chimpanzees at Assirik occupy an environment much more like that of the San or Hadza (see Table 6.1a) or Plio-Pleistocene hominids at Olduvai (Table 6.1c) or Laetoli or Hadar (Tuttle, 1988) than that of their forest or woodland dwelling counterparts at Tai or Kasoje or Gombe (Table 6.1b). Thus, any attempts to tie hominisation to habitat on the grounds of climatic differences seem simplistic.

Case study: Tasmanian humans and Tanzanian apes

It is hard enough to compare the material cultures of similar peoples, such as !Kung versus G/wi San, much less dissimilar ones, such as Inuit versus San. Drawing comparisons across species is even harder. First, what is needed is a comprehensive but precise, rich yet objective taxonomy that is neither ethno- nor anthropocentric. Several candidates exist, but the most apt typological system is that of W.H. Oswalt (1973, 1976). Second, two data-sets are needed, one human and one non-human, collected in similar ways. Such data exist for the aboriginal peoples of Tasmania and the chimpanzees of Tanzania. Third, what is wanted is an evolutionarily significant focus, a part of daily life which is undeniably subject to natural selection in terms of individual survival and reproductive success. Food-getting fills this need.

Oswalt's taxonomy

Oswalt (1976, p. vi) starts from the premise that all peoples make objects in order to obtain food, and so artefacts devoted to food production are the most crucial in any people's inventory. In the evolution of culture, subsistence technology is central. His taxonomy, or *technosystem*, is 'designed to gauge technological complexity within a single framework for the manufactures of all peoples' (1976, p. 17).

The basic structure is hierarchical and dichotomous, and the taxa are carefully defined and labelled (see Table 6.3 and Figure 6.4). Any subsistant can be classified, from the simplest naturefact to the most complex artefact. The latter may be an implement, that is an instrument or weapon, or a facility. Facilities may be tended or untended.

Such a system allows the qualitative categorisation of any subsistant, and the set of all subsistants is the food-getting tool-kit for any culture. What allows for quantitative comparison across forms and cultures is the *technounit*, the building-block of the system. The number of technounits that comprise a finished artefact is a measure of its complexity (Oswalt,

Table 6.3. *Definitions from Oswalt's (1976) taxonomy of elementary technology*

Subsistant	Extrasomatic form that is removed from a natural context or is manufactured and is applied directly to obtain food
Technounit	Integrated, physically distinct, and unique structural configuration that contributes to the form of a finished artefact
Instrument	Hand-manipulated subsistant that customarily is used to impinge on masses incapable of significant motion and is relatively harmless to the user
Weapon	Form that is handled when in use and is designed to kill or maim species capable of motion
Facility	Form that controls the movement of prey or protects it to the user's advantage. *Tended* if physical presence of user is essential for functioning; *untended* if functions in the absence of user
Naturefact	Natural form, used in place or withdrawn from a habitat, that is used without prior modification
Artefact	End product resulting from modification of a physical mass to fulfil a useful purpose
Simple	Retains same physical form before and during use
Complex	Parts change their relationship with one another when form is used

1976, p. 43). This is superior to merely counting up the types of tools used, as the sum can be divided by the total number of technounits to give an average measure of technological complexity. For example, a hafted spear has at least a shaft, a point and a binder, giving three technounits, whereas a sharpened stick used as a spear counts as only one.

Oswalt (1976, p. 199) also seeks to go beyond the product, to classify the ways in which an artefact can be made. He states four principles of production: reduction, conjunction, replication, linkage (see Table 6.4). These he presents in an evolutionary sequence, based on the premise that technological change is cumulative.

Oswalt (1973) preliminarily applied the taxonomy to 12 non-literate peoples, limiting these to societies whose economy was based exclusively on hunting, fishing and collecting foods. In an expanded and refined application he took 36 societies, having added cultivators of roots and cereal crops to the foragers (Oswalt, 1976). Bio-geographically, the chosen societies ranged from the tropics to the arctic, from deserts to forests. Others have used the taxonomy too: Lustig-Arecco (1975)

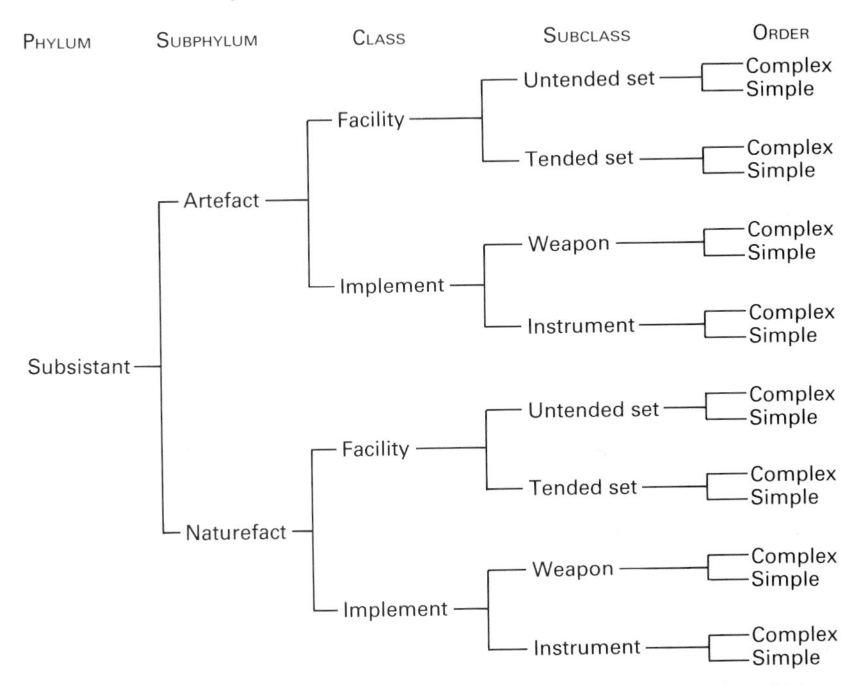

Figure 6.4. Taxonomy of subsistants, i.e. objects used in food-getting. (After Oswalt (1973, 1976).)

Table 6.4. *Oswalt's (1976) principles of production of artefacts*

Reduction	Reduce mass of form, whether natural or man-made, to produce a functioning form, e.g. flaked stone
Conjunction	Combine two or more technounits to create a finished form, e.g. hafted axe
Replication	Craft two or more similar structural units used to function as one part of a form, e.g. prongs of leister
Linkage	Use physically distinct forms in combination to perform particular purpose, e.g. bow-and-arrow

extended it to pastoralists. Torrence (1983) used it analytically to test ideas about time-budgeting and the diversity of hunter-gatherer tool-kits.

Until recently the system was explicitly presented in human terms. Non-human species, especially chimpanzees, were mentioned, but only briefly and with little optimism (Oswalt, 1973, pp. 16–17; 1976, pp. 19–20). Their use of implements was said to be uncommon, and to have little if any bearing on the development of technology among humans.

Oswalt (n.d.) has since compiled a pan-species inventory of technological forms, but my attempt to apply his taxonomy in a one-to-one comparison between two species of hominoids was done independently (McGrew, 1987).

Choosing samples

In seeking to model the transition between the pre-human and human stages of evolution, it makes sense to narrow the gap. That is, one should seek the most complex technology in the non-human species, and the simplest one from the human array. However, finding two such end-points in the two spectra is only the first step. The data to be used must also meet methodological criteria of comparability: anecdote should not be set against systematic ethnography, nor observations of behaviour against products of behaviour, nor a sample of one subject or group against multiples of these, nor brief studies against long ones.

On the main point, the Tasmanians are said to have had the simplest technology of all human foraging peoples (Jones, 1977, p. 196; 1984, p. 46; Oswalt, 1973, pp. 91–2, 96; Toth & Schick, 1986, p. 73). For example, they lacked pottery, metals and the bow-and-arrow (Plomley, 1966). They were even much impoverished in contrast to their counterparts in mainland Australia (Hiatt, 1968, p. 217; Oswalt, 1976, p. 172). They are often described in terms of what they lacked: either hafted or ground stone tools, bone tools, nets, fish-hooks, shields, spear-throwers, boomerangs, canoes, dogs and fire-making. All of these were found on the mainland to the north.

The chimpanzees of the eastern shore of Lake Tanganyika, in western Tanzania, are the best known non-human tool-users in the world. Most data come from Gombe (Goodall, 1968) or Kasoje (Nishida, 1990). They too are a relict, marginal population separated by a water barrier from a main, larger population. Thus, the main condition for comparison is satisfied.

Methodologically, our knowledge of the Tasmanian humans and Tanzanian apes is remarkably similar. In both cases the data are largely descriptive natural history. Neither set was collected by trained ethnographers. The quality of data varies from brief anecdotes of single cases to detailed accounts of repeated, first-hand observation. In both cases some data are directly behavioural while others are indirect and based only on artefacts. Observational conditions vary in both types from good (that is, close-up, friendly subjects, clearly visible) to bad (that is, long-range, shy subjects, obscured). Both sets cover both sexes and all ages from a range of communities living in a variety of habitats, so giving a regional, composite picture. Both data-sets are embedded in wider-ranging accounts of the lives of the subjects, and neither was collected with material culture as the main interest. Both sets were collected over decades by several workers, yet in each case the core of findings comes from an intensive period of a few years: G.A. Robinson (Plomley, 1966) in 1829–1834 in Tasmania, and J. Goodall (1968) in 1960–1965 in Tanzania.

Of course, there are also many differences between the two samples: the main one is that the surviving Tasmanian aborigines no longer live traditional life-styles, while the Tanzanian chimpanzees largely do. Thus, if new data are to be added, they must be recovered retrospectively for the former, while they may be planned prospectively for the latter. Further, while the human data are largely qualitative and collected by non-scientists, recent chimpanzee data are increasingly quantitative, and collected by ethologists. Offset against this, the human observational data were often accompanied by verbal explanations from Tasmanian informants, while the primatologists have to infer goals and functions from the behavioural patterns seen.

For both sets I have tried to exclude cases in which latter-day, outside influences are thought to have changed traditional practices. (In both cases, disruption of previously pristine conditions was well under way before ethnographic data collection began.) For the Tasmanians, tools and techniques introduced or altered by Europeans are left out, such as the use of spit and gloves to collect mutton-birds (Hiatt, 1968, p. 208). For the chimpanzees, applications of existing tools to newly presented human tasks are omitted, such as sticks as levers to prise open banana-feeding boxes (Goodall, 1968, p. 207). Also left out is prehistoric evidence. Although archaeological data are available for the Tasmanians (Kiernan *et al.*, 1983), these data cannot yet be distinguished for the chimpanzees from those of sympatric humans. Both of these exclusions follow Oswalt's precedents.

Lest the impression be given that the Tasmanian–Tanzanian comparison is somehow conveniently unique, it should be said that alternatives exist, for both species. On the other side of Australia the tropical Tiwi have a subsistence technology sometimes thought to be equally simple (Lustig-Arecco, 1975, p. 14; Oswalt, 1976, p. 165). For the chimpanzee, the regional population of the high forests of far western Africa, in eastern Liberia, western Ivory Coast and southern Guinea shows a lithic technology far more impressive than that of Tanzania's apes (Boesch & Boesch, 1983). Finally, it should be stressed that *both* species show more impressive forms of material culture in other, non-food-getting aspects of daily life (cf. Noble & Davidson, 1991).

Tasmanian aborigines

The Tasmanians were almost exterminated before anthropology began so we are lucky to have any ethnographic data. Robinson's extensive journals provide the bulk of the data on which later commentators have depended. According to Jones (1984, pp. 34–6), Robinson was a diligent and conscientious recorder during his long period of constant contact with the aborigines. The most complete bibliography of original sources seems to be that of Hiatt (1967).

Tasmania is a large, offshore island with a temperate, marine climate. Its vegetation ranges from rain-forest in the west to open grassland in the east. Annual burning prevents vegetational succession in the latter. The fauna is mostly Australian, but impoverished. At the time of European contact about 4000 persons in bands of 70–85 lived in territories of some 500–800 square kilometres each. Jones (1984) divided these into nine linguistic groups.

Hiatt (1967) found just under 300 observations on diet in the published literature on the Tasmanians. Overall, they were omnivorous, apparently eating about 70% animal foods and 30% plant foods (Jones, 1984). Of animals, they ate molluscs, crustaceans, birds, and mammals, especially macropods. Conspicuously absent from the diet were fish. Of plants, they ate roots, foliage, fruits, seeds, gum and fungi.

Of implements, unmodified stones were used to chop down trees to get leaves, to notch or to bruise the bark of living trees as foot-holds to make climbing easier, or to tap trunks for sap (see Table 6.5). Reeds or twisted bark were used as drinking straws to suck up sap. Stones were also thrown to knock down prey such as water-fowl. Unmodified sticks were used to dig up roots and dig out prey such as platypus. Small, wooden chisels were used to prise shellfish from underwater rocks along the shore. Also made were missile sticks to knock prey such as opossums out of trees. Kangaroos, wombats and wallabies were impaled with spears which had one end sharpened by scraping with a flint and hardened by fire.

Of facilities, spear-points were set in the ground on a kangaroo's trail, to wound an unwary marsupial travelling along it. A sort of wicker trap

Table 6.5. *Subsistants of the Tasmanian aborigines (from Plomley, 1966)*

Category	Form	Use of form	Artefact/ naturefact	No. of technounits	Page nos. in Plomley
Instrument	1. Stone	Chop down, notch, bruise, etc., living tree	N	1	188, 190, 208, 557
	2. Stick	Dig up prey	N	1	168, 544
	3. Chisel	Dislodge shellfish	A	1	63, 79
	4. Stick	Beat bushes to drive or knock down prey	N	1	162
	5. Reed	Suck up sap	A	1	534
	6. Bark	Suck up sap	A	1	534
Weapon	7. Stone	Throw to knock down prey	N	1	310, 532, 533
	8. Stick	Throw to knock down prey	A	1	162, 393, 837
	9. Spear	Stab prey	A	1	162, 379, 618
Facility, tended	10. Bark torch	Illuminate nocturnal hunt	A	1	162, 673
	11. Fire-stick	Drive or smoke out prey	A	1	837, 840, 903
	12. Rope	Climb tree to prey	A	1	190, 208, 531
	13. Grass, tied	Trip up kangaroo	A	1	218
	14. Basket	Carry shellfish	A	1	63, 79
	15. Hide (wood and branches)	Conceal hunter	A	2	559
	16. Hide, baited (sticks, grass, bait and stone)	Conceal bird-catcher	A	4	751, 813
Facility, untended	17. Spear, sunken	Wound prey on trail	A	1	626, 875
	18. Trap	Catch birds	A	1?	722, 810
				22	

1. No indication of modification, but said in one case to be 'sharp'.
4. No description given; could be same as 2.
5. Used as drinking straw.
8. Only named item in list; called *waddy*.
11. No description given; could be same as 10.
16. Satterthwait (1979, p. 413) counted worms and fish as bait as two technounits, although only one or the other was used at one time. He also included 'binders' as a technounit, although this was not mentioned in the source.
18. Satterthwait (1979, p. 414) counted as three technounits the withe framework + binders + bait, though none was mentioned in the text. Original description was too minimal for classification.

Table 6.6. *Application of Oswalt's (1973, 1976) classification of subsistence technology to the Tasmanians*

No. of subsistants	No. of technounits	Mean no. of technounits per subsistant	Source
10	17	1.7	Oswalt (1973)
11	15	1.4	Oswalt (1976)
13	17	1.3	McGrew (1987)
18	22	1.2	This study
18	25	1.4	Satterthwait (1979, 1980)

was used to catch crows, ducks, etc. All other facilities were tended: torches were used for nocturnal hunting. Fire-sticks were used to ignite grassland in order to drive kangaroos to waiting spearmen; they were also used to smoke out opossums from the hollows of trees. Also in grassland, tussocks of grass were tied together to trip up kangaroos fleeing from hunters. Grass was also plaited to make loops of rope to help in climbing trees after opossums and woven to make baskets in which to collect shellfish while diving underwater. Hides to conceal kangaroo-hunters were made of two components: a deadwood frame and a covering of branches. The most complicated facility was a hide to catch birds. A stick (pole?) supported a covering of grass, upon which a bait (e.g. fish) was 'fastened' (weighed down?) with a stone. A person hidden inside could reach through the grass to grab a bird lured down to the bait.

Attempts to apply Oswalt's taxonomy to the subsistence technology of the Tasmanians have varied in interpretation (see Table 6.6). (Satterthwait's work (1979, 1980) was unknown to me until 1989.) The number of subsistants varies from 10 to 18 and the number of technounits from 15 to 25. Regardless of which version is accepted, the conclusions given below hold true.

This tool-kit for subsistence is the simplest of all known human cultures: There were *no* complex forms of any type, and *no* compound implements, i.e. made up of more than one technounit. (Table 6.3 explains these terms.) Most subsistants (14 of 18 in Table 6.5) were artefacts, but the average number of technounits per subsistant barely exceeded 1: mean = 1.2. This simplicity is even more impressive given that it is a composite inventory for the whole island: presumably any given band had fewer tools (Oswalt, 1976, p. 175).

Tanzanian chimpanzees

The population of chimpanzees living along the eastern shore of Lake Tanganyika is split into local groups in varying degrees of isolation, sometimes far inland (Kano, 1972). Longer field-studies have concentrated on two places: Gombe and Kasoje. Shorter studies have been done at Bilenge (McGrew & Collins, 1985), Filabanga (Kano, 1971), Kabogo Point (Azuma & Toyoshima, 1961–2) and Kasakati (Izawa & Itani, 1966).

The eastern shore of the lake is defined by the escarpment of the East African rift, and in the Mahale Mountains this rises to almost 3000 metres. Rainfall varies accordingly, producing a mosaic of habitats from open grassland to evergreen forest, but the characteristic vegetation-type is deciduous, open woodland (*miombo*). The fauna is typically East African, with some additions from the Congo basin to the west.

The chimpanzees are rarely persecuted, and range from being shy and wary at Bilenge to fully tamed by provisioning at Gombe. There are well-documented differences between communities in both animal and plant foods and in tool-use to get them (McGrew, 1983; Nishida *et al.*, 1983).

Chimpanzees use slender, fragile probes to 'fish' for tree-living ants, mound-dwelling termites, and honey from both arboreal and underground hives (see Table 6.7). Similar probes are poked into inaccessible cavities for tactile or olfactory investigation, seeking prey (Goodall, 1968, p. 206). These probes are made of twigs and shoots, vines, leaves and grass-blades, stems and stalks, and strips of bark (Goodall, 1964; McGrew & Collins, 1985; Uehara, 1982). The chimpanzees use stouter wands made from branches or shoots are used to dip for terrestrial driver ants (McGrew, 1974). They crush leaves together to make a 'sponge' to wipe clean the cranial cavity of a prey (Teleki 1973, pp. 144–5) or the inside of a hard-shelled fruit (Wrangham, 1977). A similar wad of leaves is used to gather up arboreal ants from a tree-trunk for eating (Nishida, 1973). They throw stones, sticks and handfuls of leaves at baboons competing with them for food (Goodall, 1964). Once an old male threw a stone at bush pigs, apparently to force them to break ranks from their defensive formation (Plooij, 1978*b*).

All known chimpanzee facilities are tended ones (see Table 6.3). Broken-off branches are used to expel ants from their tree-nests, so that they may be caught by hand (Nishida, 1973). Sticks also are used as levers to widen the entrance of bees', termites' or birds' nests to make it

Table 6.7. *Subsistants of the Tanzanian chimpanzees (from Goodall, 1986; McGrew, 1987)*

Category	Form	Use of form	Artefact/ naturefact	No. of technounits
Instrument	1. Twig/shoot		A	2×1
	2. Leaf/grass	Fish for ants,	A	2×1
	3. Vine	honey, termites;	A	2×1
	4. Stem/stalk	investigative probe	A	2×1
	5. Bark		A	2×1
	6. Stick	Dip for driver ants	A	1
	7. Leaves	Sponge for brains, fruit pulp	A	1
	8. Leaves	Wad for ants	A	1
	9. Leaves	Brush away bees, ants	A	1
Weapon	10. Stone	Throw to drive prey	N	1
Facility, tended	11. Stick	Lever open beehive or ant nest entrance	N	1
	12. Stick	Stir up ants, bees	A	1
	13. Sapling	Elevated site for ant dipping	A	1
	14. Leafy nest	Container to crush weaver ants	A	1
	15. Leaves	Plate to catch faeces (for reingestion)	A	1
				20

1–5. Each material has double use: fishing for social insects and probing in investigation.

easier to get at the occupants (Goodall, 1986, p. 540). Saplings are bent over to make an elevated perch from which to dip for driver ants; this provides a more comfortable site away from the biting prey (McGrew, 1974). The nests of weaver ants are plucked and crushed by rolling between the palms of the hands; this *transforms* the nests into containers, killing or trapping the occupants so that they can be eaten at leisure (Goodall 1968, p. 187; unpublished data).

The chimpanzees of western Tanzania have a repertoire of 15 subsistants totalling 20 technounits; only two of the 15 are naturefacts. However, many types of tools known from elsewhere in Africa are missing: for example, brush-stick from Cameroon (Sugiyama, 1985), digging-stick from Equatorial Guinea (Jones & Sabater Pí, 1969), fruit-hook from Guinea (Sugiyama & Koman, 1979), plus the hammers and anvils cited in Chapter 1.

Table 6.8. *Comparison of production principles used by Tasmanian aborigines and Tanzanian chimpanzees in food-getting*

Oswalt's principle	Human	Chimpanzee
Reduction	Spear	Fishing probe
Conjunction	Plain hide	(Nest-building)
Replication	Tied-up grass	Leaf sponge
Linkage	Baited hide	(Ant-dipping sapling)

Subsistants compared

There are many parallels between the two tool-kits. In both cases, all subsistants are simple; none is complex or compound. The ratio of artefacts to naturefacts is much the same. Both tool-kits focus on the same raw materials: woody vegetation, stone, non-woody vegetation. Both use tools mainly for animal rather than for plant prey. Both emphasise tended rather than untended facilities. Both 'outwit' prey: for example, human hide and chimpanzee perch.

There are differences too: Only the humans use subsistants of more than one technounit. Only the humans use untended facilities and fire, and show more specific strategies such as knot-tying and baiting. On the other hand, the apes' tool-kit seems more plastic; five types of flexible probe are used for four distinct tasks. All chimpanzee artefacts are made with the hands and teeth, whereas at least some (but not all) human artefacts are made with other tools, mostly of stone (Noble & Davidson, 1991).

The two tool-kits can also be compared on Oswalt's principles of production (Table 6.4): the aborigines showed all four, while the Tanzanian chimpanzees show two or three (see Table 6.8). Tasmanian spears were made by reduction, in that a stick was sharpened at one end. Their hides showed conjunction, in that branches were arranged on a wooden frame. The tied-up tussocks of grass for tripping prey showed replication. Finally, the baited hide showed linkage, in that the hide concealed the hunter and the bait lured the prey.

For chimpanzees, the fishing-probes are reduced, for example, by twigs being torn from shrubs, the leaves stripped and bark peeled, and the ends clipped (see Figure 6.5). The leaf-sponge shows replication in that a composite, crushed mass is made from essentially identical elements. The bent-over sapling upon which the chimpanzee ant-dipper sits, plus the dipping wand used, may be linked forms (Oswalt, 1976,

Figure 6.5. Preparation for ant-dipping by adult female, Gombe, Tanzania: (*a*) stripping leaves from tool; (*b*) peeling bark from tool; (*c*) clipping tool to right length; (*d*) digging open ants' nest by hand. (Courtesy of C. Tutin.)

(c)
(d)

p. 204, personal communication). Even if the Tanzanian wand–sapling connection is not a valid linkage (T. Nishida, personal communication), chimpanzees elsewhere show linkage between stone hammer and anvil. What is missing from the apes' subsistence tool-kit is conjunction, in that no subsistant consists of combined forms, that is, comprises more than one technounit. (However, on another front, the sleeping platforms or nests built by chimpanzees each night *do* show conjunction, in that they combine broken-off branches and a lining of leafy twigs.)

As expected, the subsistence ecology of the human society is more complicated than that of the ape. However, the difference is far from wide, and the gap between hominid and pongid is bridgeable. Evolutionarily, one can imagine the subsistence technoculture of an intermediate, ancestral hominoid filling the gap. Even more intriguingly, the contrast shown here could easily be cultural, without resort to phylogenetic differences. *Given what is known of chimpanzees' abilities in captivity* (Hayes & Hayes, 1954; Brink, 1957; Beck, 1980, pp. 111–15), *they are capable of making and using all of the subsistants in Tasmanian material culture, including conjunctive production.*

Similarities and differences

Given that direct comparisons cannot yet be made between sympatric African hunter-gatherers and apes, the next best thing may be to point out potentially fruitful areas for further study. That is, tentative hypotheses based on present fragmentary knowledge can be posed, with the proviso that some of these are now little more than notions. To try to illuminate contrasts, I will focus on both similarities and differences, and make some gross generalisations about the food quest. This preoccupies both types of forager: Wiessner (1981) eavesdropped on 76 conversations lasting at least 15 minutes in a !Kung San camp, and 59% concerned the availability, procurement or redistribution of food. Wrangham (1977) found that chimpanzees at Gombe spent on average half their waking time eating, that is, feeding took up more of daily life than all other activities combined.

Diet

Foraging apes and humans show similar strong preferences for certain plant (fruit, nuts, seeds) and animal (mammals, insects, honey) foods. Their common interest in the reproductive parts of plants is not surprising, as these contain both energy and protein. Some species of fruits, such as *Pseudospondias microcarpa*, seem to be eaten by all humans and apes who can get them, right across Africa (Tanno, 1981; Isabirye-Basuta, 1988; McGrew *et al.*, 1988). Nuts are even more energy-rich, as well as nutritious, because of their high fat content (Peters, 1987*a*). Among humans, the best-known example is the mongongo nut, a sort of wonder-food for all seasons crucial to the subsistence of the !Kung San (Lee, 1968; Peters, 1987*b*). Among apes, the widespread oil palm nut discussed in Chapter 1 is the best example.

Both human and non-human foragers focus on mammals as opposed to other vertebrate classes; only birds also figure in both diets and then only opportunistically. Differences emerge, however, with habitat. Open-country hunters like the San often pursue large prey with prolonged search-and-stalk tactics; forest-living hunters like pygmies, whether archers or net-hunters, usually capture smaller prey, within the size range of chimpanzees' prey. For Mbuti net-hunters, almost 60% by number and over 40% by weight of prey came from the blue duiker, which weighs less than 5 kilograms (Ichikawa, 1983). Chimpanzees also eat this species (Nishida & Uehara, 1983), so it is likely that humans and non-humans are in direct competition for it when they live sympatrically.

Both hominoids focus on social insects, whose disadvantageous small body-size relative to a large-bodied predator is offset by their advantageous concentration in space and large collective biomass. Large, mound-building termites are preferred, and both humans (Tanno, 1981) and apes (McGrew *et al.*, 1979*a*) use ingenious techniques to overcome the prey's defence. Honey is arguably the most prized single food in the diets of both hominoid species, as both humans (Ichikawa, 1981) and chimpanzees (Brewer & McGrew, 1990) work hard, risk hazards and suffer pain when raiding the hives of honey bees.

A notable similarity of African hunter-gatherers and apes is the near total absence of the use of grass seeds (Tanno, 1981). Even in habitats dominated by grasses, such as Assirik in Senegal (McGrew *et al.*, 1981) or the Hadza country around Lake Eyasi in Tanzania (Woodburn, 1968), wild cereals are ignored. Amongst tropical hunter-gatherers world-wide there seems to be only one case of wild grains being exploited as a staple, that being *panara* by Australian aborigines (Tindale, 1977). Even so, the labour of collecting the seeds is done by ants, which are then parasitised by humans. Given the known importance of grass seeds to some living, large-bodied primates in Africa, such as the gelada baboon of Ethiopia (Wrangham, 1980*a*), and the hypothesised importance of grass-seed-eating in human evolution, based largely on dental evidence (Jolly, 1970), its absence from the diet of both types of African foraging hominoids is remarkable. Similarly absent from the diets of both humans and apes, at least to any substantial degree, are nectar and exudates.

Contrasts in diet also emerge. The most striking is that of underground storage organs, that is roots and tubers, the high carbohydrate pay-off which depends on energetically expensive excavation. Tubers provide starchy sugars for hunter-gatherers living both in forest (Tanno, 1981) and on savanna (Vincent, 1984), especially the latter. Chimpanzees only rarely eat roots, and when they do these are small bulbs simply pulled up by hand (McGrew *et al.*, 1988) or surface roots directly gnawed (Nishida & Uehara, 1983). However, wild chimpanzees *do* know how to dig with tools because they use sticks to break up termite mounds (Jones & Sabater Pí, 1969), and so they could just as well dig up roots.

Further, at least some African foragers frequently eat fungi (Mbuti pygmies: Tanno, 1981), while no chimpanzees commonly do so, though bonobos may (Kano & Mulavwa, 1984). Few African hunter-gatherers consume many of the structural parts of plants such as stems, stalks and leaves (Tanno, 1981), whereas chimpanzees throughout the species' range do so daily (Goodall, 1968; Nishida & Uehara, 1983; Sugiyama & Koman, 1987).

In summary, it can be hypothesised that the key dietary changes in hominisation were the addition of large mammals and tubers to the basic ancestral hominoid's diet. In the absence of anatomical specialisations for predation and digging, such as claws, both of these resources require technological aids for exploitation.

Food acquisition and processing

Both humans and chimpanzees scavenge meat, that is, they appropriate prey killed by other predators. For hunter-gatherers the best quantitative data are for the Hadza (Bunn *et al.*, 1988; O'Connell *et al.*, 1988). Over 20% of large (> 40 kilograms) mammal carcasses were scavenged, most of them by driving off large carnivores such as lions. Tanzanian chimpanzees pirated prey from baboons (Morris & Goodall, 1977) and stole cached or abandoned prey from carnivores (Hasegawa *et al.*, 1983). The main differences seem to be that humans regularly displace larger or more dangerous competitors from large prey, while chimpanzees occasionally take advantage of less risky opportunities to take small prey. In rain-forest habitats neither humans nor apes seem to scavenge (cf. Kortlandt, 1967).

Although both apes and hunter-gatherers show wide dietary diversity (omnivory), both also show puzzling omissions from their repertoires. Both do not eat animals and plants that seem to be edible and readily available. Mbuti pygmies avoid eating some species of birds such as francolins and mammals such as chimpanzees, and restrict the eating of others at certain stages of the human life cycle (Ichikawa, 1987). No chimpanzee has ever been known to kill and eat a reptile, even abundant and apparently accessible species such as monitor lizards. Similarly, some seemingly vulnerable species of mammals such as porcupines are not eaten by apes. Both human and non-human foragers show some similarities in dietary restrictions, in that both generally avoid eating carnivores and instead focus on common, staple species (Ichikawa, 1987).

All African hunter-gatherers use containers for acquiring, transporting and storing items, usually food. For the Mbuti, these range from the single folded leaf of a Marantaceae plant as a temporary packet, to woven hunting nets many metres long (Tanno, 1981). In contrast, wild chimpanzees use only a few natural containers such as leaf sponges and weaver ants' nests. Hunter-gatherers make few containers for food processing, but acquire items such as cooking pots from their neighbours. They also use containers to transport their subsistants (Woodburn, 1970), while chimpanzees either make and then discard tools on the spot (McGrew, 1974) or carry tools directly by hand from place to place (Boesch & Boesch, 1984*a*).

Most raw materials for tools are organic (animals and plants) rather than inorganic (stone, clay, etc.) for both human and non-human foragers (Tanno, 1981). The proportion of plant to animal raw materials varies from high (Mbuti) to low (San) for humans, but chimpanzees in nature apparently never use animal matter such as skin or bone for tools. Both hunter-gatherers and chimpanzees are limited by availability of key raw materials for subsistence tasks. Mbuti net-hunting is constrained by the abundance of the bark (*Manniophyton fulvum*) used to make their nets (Tanno, 1981). Chimpanzees' termite-fishing at Assirik is concentrated in the habitat-type where the preferred species supplying the twig tools (*Grewia lasiodiscus*) is found (McBeath & McGrew, 1982).

All hunter-gatherers use fire to cook food. However, not all hunter-gatherers make fire, many foods are eaten uncooked, and until cooking vessels were introduced in modern times cooking was probably confined to roasting on coals (Stahl, 1984). Chimpanzees being rehabilitated into the wild in Senegal spontaneously ate wild seeds parched or dehisced by bush-fires (Brewer, 1978, p. 232), but no wild chimpanzee has yet been seen to do this (McGrew, 1984, 1989*a*). Ability to *control* (as opposed to make opportunistic use of) fire is not likely to be a crucial distinction between humans and non-humans (cf. Goudsblom, 1986). Cigarette-smoking chimpanzees in the Johannesburg Zoo regularly maintained and extinguished fire in the pursuit of their addiction (Brink, 1957). Brewer's rehabilitated chimpanzees imitatively managed campfires in rudimentary ways for cooking and warmth (Brewer, 1978, pp. 174, 176). Russon & Galdikas's (1992) rehabilitated orang-utans imitated all the components of making and sustaining campfires.

All of the above topics – scavenging, dietary omissions, containers, raw materials, cooking – are ones in which there is a bridgeable gap between human and non-human forager. Hypotheses about hominisation emerge when one imagines an ancestral hominoid going beyond the 'furthest' point reached by living apes. Evolutionary reconstruction thus becomes a matter of plausible 'next steps'. It is not so easy to work from the other end backwards, that is, to imagine the last step preceding a particular subsistence technique used by a living hunter-gatherer. Use of recently acquired technology bedevils intepretation: for example, did the Mbuti *really* never boil food until they acquired cooking vessels (Tanno, 1981), and if so how did they eat plant foods which need leaching of secondary compounds in order to be edible?

What must be remembered is that many subsistence techniques and tools are uniquely human. That is, for almost every ape technique there is a human counterpart, but the reverse is not true. Only humans use guided missiles to bring down prey, whether these be boomerangs or arrows. Only humans set untended facilities, whether these be snares, traps, pitfalls, etc. Only humans use poisons, whether on tipped darts or in dammed-up streams. Only humans use dogs as hunting companions, or a mortar-and-pestle to crush plant foods. And so on. It seems likely that most of these aspects of material culture are securely hominid, that is, innovations by humans after the transition of hominisation was completed.

Conclusions

African apes and hunter-gatherers can be profitably compared, if one is seeking clues upon which to model hominisation. At the very least, such comparisons point out gaps that are small enough to investigate further. The Tasmanian human versus Tanzanian non-human exercise shows that such comparisons can go beyond the speculative to the systematic and even quantitative. To assess the extent of flexibility and variation in the hominoid half of the comparison means using similar analyses across the range of African apedom, from Uganda to Senegal. This is the aim of the next chapter.

7
Chimpanzees compared

Introduction

It should be clear by now that there is no such creature as 'The Chimpanzee'. Earlier chapters have shown enough variance in the data to make any attempt to generalise about the whole species a nonsense. However, until now the comparisons advanced have been selective or superficial. The aim of this and the following chapter is to compare systematically the object manipulation of as many populations of wild chimpanzees as possible, to see if real differences exist between them and, if so, why. If differences emerge, explanations will be sought in terms of the environmental and social contrasts which characterise those populations. Prized natural foods that require processing will be emphasised: meat, termites, ants, honey, nuts.

Of constant interest is the matter of culture, that is, whether or not one must invoke some kind of social learning of traditions or customs in explanation (McGrew & Tutin, 1978; Nishida, 1987). *In other words, do chimpanzees passively and individually react to environmental forces, or do they actively seek and acquire essential knowledge from one another?* To answer this compound question we must try to do ethnography on a non-human species, with a view to applying this in ethnological analyses (see Chapter 8).

Difficulties of comparison: eating meat

As discussed in Chapter 2, methodical comparisons across populations are most reliable when based on long-term studies, that is, those of at least a year's duration. For example, seasonal variation is crucial when considering insects as prey (see below), since many species are only available for limited periods (Janzen & Schoener, 1968). Similarly, chimpanzees' preying on mammals appears to be seasonal, at least in some places (Kawanaka, 1982; Norikoshi, 1983; Takahata *et al.*, 1984). Although there are only 11 long-term field-studies of chimpanzees (see Table 2.2), patchy data from shorter studies is also informative and so will be used when needed.

Sampling is also a problem (Martin & Bateson, 1986). Short studies may miss rare events altogether, or may distort true relative frequencies through the bias of small samples. For behavioural data there is great variation across field-sites in the quality of opportunities available to investigators. Unless all age- and sex-classes of subjects are equally observable, apparent differences within and across populations may be spurious. These problems are compounded by provisioning, which too has biasses, and only Tai's apes have so far yielded detailed behavioural data without any provisioning (Boesch & Boesch, 1989).

The following example shows the methodological problems of comparative studies by focussing on a deceptively simple question: *How often do chimpanzees eat meat*?

Of the classes of vertebrates, only mammals regularly fall prey to chimpanzees (see Chapter 5). Over the collective geographical range of the apes, at least 25 species of mammals are taken, varying in size from mice to juvenile bush pigs (Goodall, 1986; Wrangham & van Zinnicq Bergman Riss, 1990).

The widest variety of answers to the question of frequency of meat-eating comes from Gombe. The most *extensive* analysis was by Wrangham & van Zinnicq Bergman Riss (1990), whose data-set comprised 14 583 hours of focal-subject data collected over 4 years on two neighbouring communities of chimpanzees. The animals averaged 200 kills totalling more than 600 kilograms of meat consumed per year, at an overall rate of 0.18 kills per 100 hours of observation. The most *intensive* study was by Riss & Busse (1977) who compiled a 50-day continuous record of the waking life of an adult male chimpanzee. Over this period, totalling 563 hours of focal observation, he made three kills and twice ate meat killed by others. Of the *non-behavioural* data, the most standardised and easily collected are faecal specimens in which presence or absence of remnants of prey are recorded. Table 7.1 gives details for Gombe: in 42 months of data collected by Goodall, 5.8% of specimens contained remains of mammals.

The next most studied population of chimpanzees is that of Kasoje. Data were not available from focal sampling, so findings were opportunistic (*ad libitum*). Over 3 years, eight kills occurred in 1415 hours of

Table 7.1. *Rates of meat-eating by various populations of chimpanzees, as measured by analyses of faecal specimens*

Site of study	Length of study (months)	Total specimens (A)	Specimens with mammalian remains (B)[a]	B/A (%)	Source
Gombe[b]	42	1963	114	5.81	J. Goodall (unpublished data)
Kasoje	83	4217	48	1.14	Takahata *et al.* (1984)
Tai	?	381	1	0.26	Boesch & Boesch (1989)
Assirik	43	783	14	1.79	McGrew (1983)
Kasakati	15	174	1	0.57	Suzuki (1966)
Bossou	6?	300+	0	0.00	Sugiyama & Koman (1987)

[a] Kasoje and Tai report specimens with *vertebrate* remains.
[b] McGrew *et al.* (1979*b*) erroneously reported a lower rate of meat-eating for Gombe, based on a mis-reading of Goodall (1968, p. 184). The rate reported here represents the total data-set from June 1964 to December 1967.

unsystematic observation, giving a rate of one kill per 177 hours (Nishida *et al.*, 1979). Predatory 'episodes', that is, all evidence of predation pooled, including unsuccessful attempts, were apparently more common, with 54 occurring in 34 months (Takahata *et al.*, 1984). The *only* measure that can be directly compared with that at Gombe is from faecal sampling (see Table 7.1): 1.1% of samples collected over 8 years contained *vertebrate* (that is, mammal and bird) remains (Takahata *et al.*, 1984).

At Tai, Boesch & Boesch (1989) presented data on hunts and kills by chimpanzees living in dense evergreen forest. They focussed on 2 years of behavioural data on habituated subjects in an effort to minimise bias, but the sampling method used is unclear. Hunts (N = 100) occurred about every 3 days on average, and 57% of these were successful, that is, at least one prey, usually a monkey, was killed. However, faecal data yielded only the bones of a bird and an overall percentage of vertebrate remains in < 1% of samples (see Table 7.1).

For the other long-term sites, only anecdotal and descriptive behavioural data are available. At Assirik, no kills were seen, but remains of mammals turned up in 1.8% of faecal specimens (see Table 7.1; also McGrew, 1983). At Kasakati, one kill, of a red-tailed monkey, was seen (Kawabe, 1966), but a small series of 174 faecal samples yielded only one with mammalian remnants (see Table 7.1; also Suzuki, 1966). At Bossou, five cases of predation on tree pangolins have been reported by Sugiyama & Koman (1987), but faecal analysis has yet to yield any remnants of animal prey (see Table 7.1).

At Budongo, three kills, including one case of cannibalism, were seen in identified chimpanzees over 17 months of opportunistic study, but no faecal data were reported (Suzuki, 1971). In a 22-month-long study at Kibale, Ghiglieri (1984, p. 72) saw one incident of meat-eating by partly habituated chimpanzees, but took no faecal data. At Lopé, C. Tutin *et al.* (personal communication) have seen predations and found remains of prey in faeces of chimpanzees, but these data remain to be analysed. Only at Okorobikó and at Kabogo have long-term studies of wild chimpanzees failed to yield any evidence of meat-eating, but the methods used have not been fully elucidated (Azuma & Toyoshima, 1961–2; Jones & Sabater Pí, 1971).

So, how often *do* chimpanzees across Africa eat meat? The above data show no clear answer. Depending on the measure chosen, *most* of the 11 sites could be ranked from first to last in frequency! Even for Okorobikó and Kabogo, it could merely be a case of absence of evidence. Furthermore, there are apparent internal inconsistencies: Tai would seem to rank first on frequency of hunts but almost last in terms of faecal remnants. The most discouraging aspect is the minimal comparability of indicators used. Only presence or absence in faeces comes close to being a standard, being available from six of 11 sites, but even this is somewhat compromised because two of the six give data on vertebrates while the other four use mammals.

The reluctant conclusion is that empirical comparisons across populations of chimpanzees, even on seemingly simple and straightforward points, must be done with caution, and the aspiring ethnologist must be prepared for frustration in doing so.

Chimpanzee insectivory

Studies of chimpanzees eating insects have concentrated more on the methods used to obtain them than on the prey taken. The methods used (see below) may yield useful clues to the mental abilities of apes (as discussed in Chapter 3), but this neglects the basic ecological significance of insects in the diet.

Two elementary points illustrate this. First, insect-eating by chimpanzees is more common than meat-eating, although the latter has

Table 7.2. *Overall consumption of animal prey by wild chimpanzees based on faecal analysis*

	Field-site			
Type of prey	Assirik	Gombe	Kasoje	Sources
Vertebrates	2%[a]	6%	1%	Baldwin (1979), McGrew (1983), Takahata *et al.* (1984), J. Goodall (unpublished data)
Termites	27%	15%	2%	Baldwin (1979), McGrew (1979, 1983), Uehara (1982)
Ants[b]	24%	22%	23%	Baldwin (1979), McGrew (1979, 1983), Nishida & Hiraiwa (1982)
Bees	23%	3%	1%	Baldwin (1979), McGrew (1979, 1983), Nishida & Hiraiwa (1982)

[a] Values are percentage of faecal specimens containing remnants of prey.
[b] Minimal figure.

claimed much more attention from both investigators and commentators (Butynski, 1982) (see Table 7.2). Insect-eating is virtually a daily activity, while meat-eating may occur only a few times a year, for a given individual. Nutritionally, eating insects is likely to be more important, on both a day-to-day and an overall basis, either in terms of nutrients or calories. Thus, it deserves comparative scrutiny. Second, unlike most other primates, chimpanzees focus their predatory activities on *social* insects, such as termites and the colony-living forms of ants and bees. (Chimpanzees also take non-social forms, such as gall insects, but this usually occurs in the course of general foraging, without special techniques.) The advantages and disadvantages are clear: Social insects are a sizeable, concentrated, and often sedentary biomass. Exploiting them is energetically efficient for a large-bodied primate. Immature forms such as larvae, which are nutritionally richer, can be taken at the same time as adults (Redford & Dorea, 1984). On the other hand, social insects are formidable in defence, having either venomous stings and painful bites or substantial and relatively impregnable homes. Among the non-human primates, only chimpanzees have solved the daunting challenges of the bee hive and the termite mound.

In comparing insect-eating here, prey are considered at the generic level only. This avoids excessive detail, but also the particular species taken is usually not known. It seems likely that from the apes' point of view major differences do not exist between species of the same genus of prey (cf. Boesch & Boesch, 1989). Table 7.3 lists genera of social insects eaten by chimpanzees at 10 sites of long-term study.

Conveniently, different types of social insects eaten by chimpanzees seem to have different roles in the diet. Termites and ants seem to be *staples*, that is mainstays of animal matter in the diet in terms of frequency or volume of consumption. On the other hand, honey qualifies as a *treat*, that is, a food-stuff of high quality and much sought after even if its contribution to overall intake is minimal.

Termites

Chimpanzees eat several species of termites, but by far the greatest numbers are of the genus *Macrotermes*, a mound-building form that farms underground fungus-gardens. The genus is impressive in all ways: its individuals are the biggest in size in Africa; a single mound may contain 2 million members; it is distributed throughout sub-Saharan Africa in a wide variety of habitats (Howse, 1970). (It's no wonder these insects have been nicknamed Big Macs!) Accessibility to termites varies with season: where there is a pronounced dry season they retreat during this period to safety far underground, and their mounds bake hard in the sun. The only other genus reported in the diet of chimpanzees at more than one site is *Pseudacanthotermes*, which is also a mound-building fungus-grower.

Gombe provides the best observational data on the consumption of *Macrotermes* by chimpanzees (Goodall, 1968; McGrew, 1979). Chimpanzees eat them in all months of the year, but consumption peaks sharply at the start of the rainy season, when female chimpanzees spend up to 15% of their waking hours termite-fishing (see Figure 5.2). Data from observations and faecal specimens agree closely. Goodall (1968) also reported that *Pseudacanthotermes* were eaten during 2 months, but this was seen fewer than 20 times, and only winged reproductive forms were taken by hand (Wrangham, 1975).

For Kasoje, Nishida & Uehara (1980) hypothesised that *Macrotermes* was uncommon within the home-ranges of their main study-groups, K and M. However, they cited convincing circumstantial evidence of tool-use by chimpanzees in a neighbouring study-group, B. McGrew & Collins (1985) confirmed that B-group's apes fish for and eat *Macrotermes*. For *Pseudacanthotermes*, Uehara (1982) found that K-group ate these termites occasionally (see Table 7.2) by toppling the towers of the mounds by hand or rarely by fishing with tools. The insects appeared in

Table 7.3. *Social insects eaten by wild chimpanzees at sites of long-term study*

	Site									
Type of prey	Assirik	Bossou	Budongo	Gombe	Kasakati	Kasoje	Kibale	Lopé	Okorobikó	Tai
Termites										
Macrotermes	+/+	+/?	+/?	+/+	+?/?	+/+[a]	+?/+	+/−	+/+	+/?
Pseudacanthotermes	−/−	?/?	+?/?	+/+	?/?	+/+	?/?	?/?	?/?	−/−
Ants										
Camponotus	+/+	?+/?	?+/?	+/−	+/+	+/+	?+/?	+/+	?+/?	?+/?
Crematogaster	−?/−?	?/?	?/?	+/+	+/?	+/+	?/?	+/?	?/?	?/?
Dorylus	+/+	+/+	+?/?	+/+	+/?	+/−	+/?	+/−	?/?	+/+
Megaponera	+/+	?/?	+?/?	+/−	+/?	+/−	+?/?	?/?	?/?	?/?
Monamorium	?/?	?/?	?/?	?/?	?/?	+/+	?/?	?/?	?/?	?/?
Oecophylla	+/+	+?/?	+?/?	+/+	+/+	+/+	+?/?	+/+	+?/?	+?/?
Tetramorium	?/?	?/?	?/?	?/?	?/?	+/+	?/?	?/?	?/?	?/?
Bees										
Apis	+/+	+?/?	+/?	+/+	+/?	+/+	+?/?	+/+	+?/?	+/+
Trigona	?/?	?/?	?/?	+/+	+/+	+/+	+?/?	+/+	?/?	+/+
Xylocopa[b]	?/?	+/+	?/?	?/?	?/?	+/+	?/?	?/?	?/?	+/+

+, present/eaten; −, absent/not eaten; +?, probably present/probably eaten; −?, probably absent/probably not eaten; ?, unknown.

[a] +/+ for B-group but virtually −/− for K- and M-groups.
[b] Some species of *Xylocopa* are solitary and others minimally social (Anzenberger, 1977).

faecal specimens in 8 months of the year, with the highest monthly figure being 6.5%.

At Assirik, *Macrotermes* was the only kind of termite found in faecal samples, and it was the most common species of insect in the diet. Consumption showed marked seasonality, and in the peak months over half of the samples contained termites' remains. *Pseudacanthotermes* was absent at Assirik, but other genera of termites such as *Cubitermes* were commonly found.

At Tai, chimpanzees ate five species of termites (unspecified) but none involved the use of tools (Boesch & Boesch, 1990). Similarly, translocated chimpanzees at Ipassa Reserve in north-eastern Gabon ate several species of smaller termites by hand but ignored the abundant *Macrotermes* (Hladik, 1973).

At Okorobikó, chimpanzees were not seen to eat termites, nor were faecal data presented (Jones & Sabater Pí, 1971), but many tools were found at *Macrotermes* mounds (Jones & Sabater Pí, 1969). Sabater Pí (1974) once saw chimpanzees using sticks to break open a mound, presumably to obtain the occupants. Similar perforating or digging sticks were found at *Macrotermes* mounds at Belinga in north-eastern Gabon (McGrew & Rogers, 1983).

A variation of the probing stick is the brush-stick, first described at Campo in Cameroon by Sugiyama (1985). The end of the tool inserted into the mound was frayed to resemble a paint-brush, apparently to increase its 'affixibility' to the biting insect defenders. Similar brush-sticks were used by the chimpanzees of Congo and Central African Republic (Fay & Carroll, 1992). What is not yet clear is whether such fraying was done deliberately by pounding with a hammer-stone (Sugiyama, 1985) or chewing with the molars (Fay & Carroll, 1992), or was an inadvertent by-product of wear through repeated use (McGrew & Collins, 1985). Behavioural data are needed.

At Kasakati, little is known about termite-eating. Suzuki (1966) reported a single case of chimpanzees being found fishing at a termite mound. His analysis of faecal specimens showed that 2% contained termites, but in neither data-set were the prey identified. At Kibale, Ghiglieri (1984, p. 72; 1988, pp. 121–2) once saw a chimpanzee eat unspecified termites by hand from a rotten log. At Bossou, chimpanzees were once seen to use tools differently to exploit unidentified arboreal termites (Sugiyama & Koman, 1979). They jammed twigs into tree-holes, squashing the insects on the tool's tip, from which the apes licked them off. However, they seemed to ignore *Macrotermes*, whose mounds were abundant.

Table 7.4. *Summary of data on free-ranging chimpanzees eating termites, with and without tools, across Africa*

Geographical race	*Macrotermes*[a]	Other termites[a]	Sources
Eastern			
Budongo	–	–	Reynolds & Reynolds (1965)
Gombe	F	H	Goodall (1968)
Kabogo	–	?	Azuma & Toyoshima (1961–2)
Kasakati	F?	–	Suzuki (1966)
Kasoje (B)	F	–	McGrew & Collins (1985), Nishida & Uehara (1980)
Kasoje (K, M)	–	F, H	Uehara (1982)
Kibale	–	H?	Ghiglieri (1984, 1988)
Central-western			
Belinga[b]	B/F/P	–	McGrew & Rogers (1983)
Campo[b]	B	–	Sugiyama (1985)
Ipassa[b]	–	H	Hladik (1973)
Lopé	–	–	C. Tutin *et al.* (unpublished data)
Ndakan[b]	B/F/P	–	J. Fay (unpublished data)
Okorobikó	P	–	Jones & Sabater Pí (1969, 1971)
Western			
Assirik	F	–	McGrew *et al.* (1979*a*)
Bossou	–	T	Sugiyama & Koman (1979, 1987)
Tai	–	H	Boesch & Boesch (1990)

[a] B, brush-stick; F, fishing probe; H, by hand only; P, perforating pick; T, other tool-type; –, not eaten or known.
[b] Short-term studies. All other studies listed are long-term.

At Budongo, no evidence of chimpanzees eating termites has been reported, although at least *Macrotermes* is present (Pomeroy, 1977). Reynolds & Reynolds (1965) found no evidence of termite mounds being disturbed in their short-term study, and later studies by Sugiyama and by Suzuki yielded no further signs of termites being eaten.

Table 7.4 summarises findings on termite-eating by free-ranging chimpanzees across Africa. Overall, chimpanzees seem to prefer *Macrotermes* (see also Table 7.3). There is an apparent positive correlation between the quality of the data and the importance of this genus of termites in the diet, given that the insects were present. Put another way, apart from Tai, no study which has produced extensive behavioural or faecal data on chimpanzees' eating habits has failed to record *Macrotermes* as a staple food, if the termites were there to be exploited. Further, whenever *Macrotermes* was consumed, by any of the three geographical races of chimpanzees, tools were *always* used. In contrast, most taking of other types of termites was by hand.

For *Pseudacanthotermes* the picture is more intriguing. Given that both Gombe's and Kasoje's chimpanzees used the same technique for termite-fishing, why did the former ignore *Pseudacanthotermes* as a fishable prey when the latter fished for it? After all, Gombe's chimpanzees fished for *Macrotermes* and took *Pseudacanthotermes* by hand, so why not fish for both? This notable omission is convincing, given 30 years of negative evidence from Gombe, and it suggests that more than environmental determinism is needed to explain the absence of fishing *Pseudacanthotermes* in the Gombe tool-kit.

Ants

Of the many kinds of ants, wild chimpanzees are known to eat only seven genera (see Table 7.3). All of these are large in size or easy to obtain, or both. No species that would require digging up from permanent or deep underground nests is taken by the apes.

Weaver ants (*Oecophylla longinoda*) live in arboreal nests made of living leaves bound together by larval silk (Holldobler & Wilson, 1977). Disturbance of a nest causes active, massed defence by painful biting. Only a few hundred ants occupy each leafy bundle, but these are usefully contained for processing, as described in Chapter 5. This is the most commonly eaten species of ant across the various populations of chimpanzees. At Assirik they were found in 24% of all faecal samples, being easily recognised by their reddish-brown heads and tiny black eyes. Their consumption by the chimpanzees was concentrated at the start of the rainy season (Baldwin, 1979). At Gombe they appeared in 22% of faecal samples, making them easily the most frequently eaten species of insect (McGrew, 1979). Wild-born chimpanzees being rehabilitated onto an offshore island in Liberia relished weaver ants, eating them more often than all other insects combined (Hannah, 1989). Weaver ants are also commonly eaten by wild chimpanzees at Lopé (D. Wroggeman, personal communication).

Elsewhere, weaver ants were less important. At Kasoje, they were found in only 2% of faecal specimens, placing them a distant fourth among species of insects eaten.

One case of weaver ant eating was seen at Kasakati (Suzuki, 1966),

Figure 7.1. Driver ants (*Dorylus nigricans*) in migration.

but they were not mentioned in reports from Bossou, Budongo, Kabogo, Kibale or Okorobikó.

Driver ants (*Dorylus* spp.) are aggressive predators that move in densely packed, branching columns on the ground (Gotwald, 1974) (see Figure 7.1). These living streams number several million members and function in omnivorous foraging and in migration. Driver ants have no permanent base, but move from one bivouac to another, building temporary underground nests. They are easily seen when moving across open ground and react to being disturbed with ferocious, biting attack. They are widely found across equatorial Africa.

As described in Chapter 5, chimpanzees use a specialised technique of tool-use – ant-dipping – to obtain this species of prey. However, each bout of dipping yielded only about 20 grams of ants (McGrew, 1974). At Gombe, driver ants were found in only about 3% of faecal samples, mostly in the wet season (McGrew, 1979). At Assirik, chimpanzees used the same technique to obtain the ants, and they turned up in only 2% of faecal specimens (Baldwin, 1979). At Bossou (Sugiyama, 1989) and at Tai (Boesch & Boesch, 1990), chimpanzees both dipped with tools and dug by hand to get driver ants.

Driver ants are present (Kabogo, Kasakati, Kibale, Lopé) or probably present (Budongo, Okorobikó) at other sites, but published accounts of the chimpanzees rarely mention them. Most notably of all, driver ants are commonly found at Kasoje but have never been recorded as being eaten by chimpanzees in 25 years of study (Nishida, 1987).

Camponotus ants are a cosmopolitan arboreal form that lives in small colonies in cavities bored from the boles of trees (Carroll, 1979). Nishida (1973) and Nishida & Hiraiwa (1982) described in detail the technique used by Kasoje's chimpanzees to obtain these ants. It resembles termite-fishing in its delicacy, but is more elaborate (see cover photograph). At Kasoje, the apes ate these ants throughout the year, and they were found in 8% of faecal specimens (Nishida, 1977). Amounts eaten were small, and Nishida & Hiraiwa (1982) characterised the consumption as virtually non-nutritional, suggesting that it may be the fiery taste which is attractive.

At Kasakati, *Camponotus* ants appeared in 4% of faecal samples, but it is not known how they were obtained (Suzuki, 1966). At Assirik, circumstantial evidence existed for chimpanzees eating these ants. Three times, ants and freshly made probes were found on the ground below *Camponotus* nests after chimpanzees had passed through the area. At Gombe, the ants occur but chimpanzees have not been seen to eat them. No data on the presence or absence of these ants, nor on whether or not they are exploited, are available from Bossou, Budongo, Kabogo, Kibale or Okorobikó.

Crematogaster ants are smaller in size and live in colonies of about 1000 in selected species of trees having a soft pith (Duviard & Segeren, 1974). They excavate a tubular hollow in the central section of a branch of suitable size. Chimpanzees do not need tools to extract them; instead they simply snap off the branch and then split it length-wise with teeth and hands. At Kasoje, chimpanzees ate these ants almost daily, in bouts of up to an hour (Nishida & Hiraiwa, 1982). They were the most frequently eaten species of insect there, being found in 28% of faecal specimens (Nishida, 1977). The only other long-term study to record chimpanzees eating *Crematogaster* was at Gombe, but it was seen fewer than 20 times (Wrangham, 1975). Apart from Gombe and Kasoje, there are no records of these ants being seen to be eaten or found in chimpanzees' faeces, nor have other investigators even established their presence at a study-site.

Megaponera is a large, dimorphic ant of pan-African distribution which specialises in raiding the mounds of fungus-growing termites (Longhurst *et al.*, 1978). After scouts find termites, 300–400 ants in a compact raiding party move briskly to the nest of the prey, seize them, and return to the temporary home-base. Emigration columns move similarly, carrying larvae and pupae. The ants possess an intensely painful sting and are quick to attack anything disturbing a column. Only the chimpanzees of Assirik are known to prey upon these ants, which occurred in 4% of faecal specimens (Baldwin, 1979). *Megaponera* occurs commonly at Gombe, Kabogo, Kasakati and Kasoje, but is not eaten by the apes. Data on the availability of the ants at Bossou, Budongo, Kibale, Okorobikó and Tai are lacking.

Several gross differences in ant-eating across populations of chimpanzees are therefore evident; species of ants which are readily available are eaten by some apes and ignored by others. Notable are: absence of *Dorylus* and *Megaponera* in the diet at Kasoje; absence of *Camponotus* and *Megaponera* in the diet at Gombe; and apparent absence of *Crematogaster* in the diet at Assirik.

Honey

Honey is the purest and most concentrated form of energy in nature that is suitable for large-bodied vertebrate predators (Fletcher, 1978). However, it is not generally appreciated that honey-bee (*Apis*) combs also contain useful amounts of protein and fat in the form of larvae, pupae and pollen. Further, when chimpanzees pillage a hive, typically by using a smash-and-grab technique, many adult bees are also caught up and consumed. All in all, a bees' nest provides a meal in itself, so it is not surprising that chimpanzees will suffer much discomfort to exploit even the fierce stinging honey-bees.

Chimpanzees also prey upon other, smaller forms of honey- or pollen-storing bees, such as the stingless *Trigona* and the solitary *Xylocopa* (Anzenberger, 1977), but no quantitative data have yet been presented (see Table 7.3). Also, with one exception, chimpanzees ignore altogether the non-food-storing wasps and hornets, which probably represent too much cost for too little benefit compared with the wingless ants. The exception is the tiny fig wasp, which is eaten inadvertently during consumption of fig fruits (Janzen, 1979).

At Assirik, honey-bees were eaten throughout the year, and were the third most common species of insect in the chimpanzees' diet (Baldwin,

1979). Tools were apparently used as probes to extract the honey (McGrew *et al.*, unpublished data; Bermejo *et al.*, 1989). At Gombe, it is unclear how often bees are eaten: fewer than 3% of faecal specimens contained bees (McGrew, 1979). Goodall (1968) reported only once seeing the eating of honey, but Wrangham (1975) listed many observations of it. At Kasoje, honey-bees rarely occurred in faecal samples (Uehara, 1986). At Lopé, chimpanzees eat the honey of both honey-bees and of other, stingless types (D. Wroggeman, personal communication).

There are no published records of chimpanzees eating honey at Budongo, Kabogo or Okorobikó, although honey-bees are presumably present at all sites, given their pan-African distribution. Only Hladik (1973) specifically noted that his translocated chimpanzees at Ipassa ignored the honey-bees that were there.

For other kinds of bees, all data on honey-eating are scattered and descriptive. The first record (Merfield & Miller, 1956) is typical: a group of chimpanzees in Cameroon used dip-stick probes to extract honey from an underground bees' nest. Gombe's chimpanzees did the same (McGrew, unpublished data). Izawa & Itani (1966) saw the same technique applied to an arboreal nest of *Trigona* bees at Kasakati, as did Fay & Carroll (1992) in Congo and Central African Republic. At Kasoje, chimpanzees use probes to get honey or larvae from both *Xylocopa* and *Trigona* (Nishida & Uehara, 1983).

Other techniques are used too. At Belinga in north-eastern Gabon, Tutin & Fernandez (1985) saw chimpanzees using only a finger to extract honey from a *Trigona* nest in a dead tree. Chimpanzees at Bossou twice removed by hand the pollen 'bread' and larvae of *Xylocopa* from a hollow tree (Sugiyama & Koman, 1987). In north-eastern Zaïre, both Goodall (1979) and Yamagiwa *et al.* (1988) found tools used by apes to dig up underground bees' nests. Fay & Carroll (1992) saw chimpanzees using wooden hammers up to 10 centimetres thick to pound open the propolis nests of meliponine bees in trees and in the ground. In the Ituri Forest of Zaïre, J. Hart (personal communication) found similar sticks used as clubs to break into bees' nests in hollow trees. At Tai, chimpanzees used tools to extract honey from four species of bees, but these have not yet been identified (Boesch & Boesch, 1990). Finally, Kibale's chimpanzees also use sticks to obtain honey (R. Wrangham, personal communication).

The most impressive technological solution to the honey-getting problem was shown by rehabilitated chimpanzees on Baboon Island in The

Figure 7.2. Tool-set used by adult female to dip for honey, Baboon Island, The Gambia: (*a*) stout chisel; (*b*) fine chisel; (*c*) bodkin; (*d*) dip-stick. (Courtesy of S. Marsden.)

Gambia (Brewer & McGrew, 1990). A female used a tool-set of four components (stout chisel, fine chisel, bodkin, dip-stick) in sequence to extract honey from a stingless bees' (probably *Trigona ruspolii*) nest in a hollow tree. She first tried to reach the honey by dipping directly from a

Figure 7.2. *cont.*

flight entrance, but when this failed she attacked the involucrum of the nest with a stout chisel. Having started an indentation, she continued to work on it with a finer-pointed chisel. This eventually allowed her to puncture the nest wall with a sharp-pointed bodkin, and then a longer, flexible dip-stick was used to dip out the dripping honey (see Figure 7.2).

In summary, chimpanzees use a wide variety of tools and techniques to get honey. It is the most widespread tool-use pattern reported for wild chimpanzees, and tools seem to be more important in honey-getting than in any other insect-eating. *All* known habitual techniques (see below) involve tools, unlike techniques for termites or ants. Also, the only known use of a tool-set in insect-eating, though admittedly only a single case, was in the pursuit of honey.

Explaining variation

How can this variation in the insectivory of wild chimpanzees best be explained? Can differences in diet or technique between populations of apes be understood only in terms of the environment or must social factors also be invoked? In the former case, differences could result from biotic or physical factors acting directly in transaction with the individual chimpanzee, without needing to posit any social learning (Galef, 1976; see also Chapter 4). In the latter case, when differences between habitats cannot account for dietary differences, then by exclusion it seems likely that different groups of apes have developed social traditions (cf. Nishida, 1987). The choice cannot be clear-cut, for we can never rule out entirely unknown (to us) environmental factors. Likewise, it is hard to show in the field that higher-order mental processes like imitation are operating in social learning (Tomasello, 1990; but cf. Russon & Galdikas, 1992). Finally, any attempt to dichotomise environmental and social factors *must* be fruitless, as each affects the other.

Clear-cut environmental determinism is most obvious with the presence or absence of a species of prey. Assirik's chimpanzees never meet *Pseudacanthotermes* termites, almost certainly because the climate of the savanna habitat is too dry for these termites, and so the apes cannot prey on them. However, because primatologists are rarely entomologically sophisticated, such conclusions must usually be drawn with caution, unless they are lucky enough to be working in an area with a well-known insect fauna.

A variant on this theme exists when a readily available species of prey is eaten by one group of apes but not by another for whom it is virtually, if not entirely, absent. Virtual absence occurs either because the prey are few or because their range overlaps only minimally with that of the chimpanzee predators. Again, lack of information about the relative abundance and distribution of insects is usually lacking, and this prevents firm conclusions being drawn. Consider the differing use of *Macrotermes* termites by various groups of chimpanzees in the Mahale Mountains (Nishida & Uehara, 1980; Uehara, 1982; Collins & McGrew, 1985, 1987). There are ecologically important differences such as rainfall between the range of B-group versus those of K- and M-groups; these correlate with differences in termite availability.

Also directly determined by the environment is the *range* of types of prey available. At Assirik, unlike Gombe, Kasoje, and probably all other forested sites, there is only one large mound-building form of termite: *Macrotermes subhyalinus*. Thus, the chimpanzees have no choice. Conversely, at Kasoje, chimpanzees of B-group ignore edible and fishable but unpalatable *Odontotermes* termites, apparently because equally accessible but bigger and tastier *Macrotermes* are available (Collins & McGrew, 1985, 1987).

Chimpanzees as predators are also constrained by the extent and nature of *competitors*, that is, by other insect-eaters trying to exploit the same species of prey. Little is known about their competitors for insects, as none except other primates has ever been studied at any chimpanzee field-site. Many species of primates eat many kinds of insects, but these are rarely social insects. Even when chimpanzees and baboons eat the same species of prey, as with *Macrotermes* at Gombe, chimpanzees are technically advantaged by their tool-use, whereas baboons eat only the rarely available winged reproductive forms (Beck, 1974). It seems likely that the chimpanzees' use of tools for getting insects minimises competition with other insect-eaters. Most techniques used by chimpanzees 'tap' but do not destroy the resource, unlike the destructive digging of honey-badgers or aardvarks.

Should any of the differences remaining after factoring out obvious environmental influences be classed as cultural? It is always risky to argue by exclusion, and it would be better to show social tradition in operation. None of the feeding habits described here meet all the operational criteria for culture set out in Chapter 4. For example, no innovation in eating a new type of natural prey has been recognised in any wild population of chimpanzees. For animal prey, the closest case may be that of the Kasoje chimpanzees eating wart-hogs (T. Nishida, personal communication). In recent years after the mass emigration of local people from the area, species of potential prey such as wart-hogs and predators such as lions have re-colonised it. Whether the advent of eating wart-hogs is truly new or just the re-emergence of an old habit cannot be determined. Lack of dietary adventurousness is not surprising, since secondary compounds in plants and venoms in animals present formidable detoxification problems. Such conservatism may be expressed in terms of a restricted 'searching image' (Krebs, 1973), a phenomenon in which a predator fails to perceive another species as a potential prey, though it may be eaten elsewhere. Boesch & Boesch (1989) reported the puzzling failure of Tai chimpanzees to eat a blue duiker fawn.

As for tradition, persistence in food-habits is clear from the long-term records of more than two or three decades from Kasoje and Gombe. But persistence is only a necessary but not a sufficient condition for tradition. Well-adapted organisms regularly cope with their environment, but do so presumably on a largely individual basis. To show a social component in the continuity of a population's diet in a constant environment therefore requires recognised innovation. This will be discussed below.

Case study: 'Fishing' for termites

However instructive, the comparison presented in the previous section on insectivory and tool-use across populations of chimpanzees is only partly satisfactory. It was largely qualitative (the data are mostly nominal or ordinal rather than interval level), superficial (even basic independent variables like tool design or raw materials are ignored), and messy (lots of confounded or missing variables prevent direct comparisons across several groups of apes). This frustration is to be expected, given the earlier example of meat-eating, but it is not inevitable.

This section aims to give an example of point-by-point comparison of an important pattern of chimpanzees' tool-use: probes inserted into termites' mounds to fish out the occupants. The data are part of a wider-ranging study (McGrew *et al.*, 1979*a*), but here the focus is on only two study-sites: Gombe and Assirik.

On gross, qualitative grounds, the two populations of apes seem to show identical technology, despite the fact that they live over 5000 kilometres apart, at opposite ends of the distribution of the species. Given data on these two populations alone, it would be easy to conclude that 'The Chimpanzee' was a termite-fisher, and perhaps just as stereotyped in this behavioural pattern as in nest-building (Baldwin *et al.*, 1981).

Consider the similarities. Both populations make and use slender probes of vegetation. Both make these from the same sort of raw

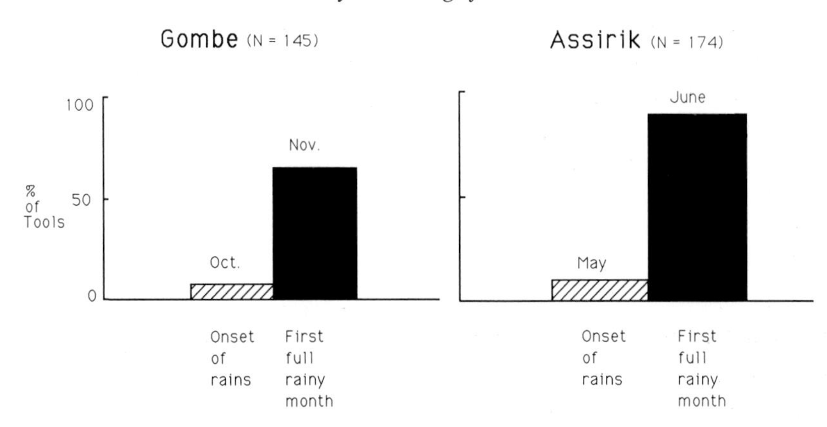

Figure 7.3. Termite-fishing tools found at termite mounds, in relation to onset of rainy season, Gombe, Tanzania, and Assirik, Senegal.

materials: twig, vine, grass. Both use the tools to extract the soldiers and workers of *Macrotermes* from within their earthen mounds. Both schedule the task at the start of the rainy season when the insects open holes in the mound's surface that allow ingress. The number of termite-fishing tools on mounds increases nine-fold from the month marking the onset of the rains to the first full month of the rainy season, at both sites (see Figure 7.3). Both consume during the season massive amounts of these otherwise unavailable subterranean food-items. Put another way, a chimpanzee from Gombe translocated in June to Assirik could walk straight to a *Macrotermes* mound and easily secure lunch.

However, it is possible to be more precise, with some quantitative similarities.

The average dimensions of fishing probes at Gombe and Assirik were much the same, as was the range (see Table 7.5). The proportion of tools that originated from raw material sources found within easy reaching distance of the mound (<2 metres) was also similar at both sites: 85% at Gombe versus 94% at Assirik.

More informative are the *contrasts* between the two sites. At Gombe, almost half of the fishing tools were made from grass blades or stems, and these plus strips of bark and segments of vine accounted for over 85% of the tools found (see Figure 7.4). At Assirik, almost half of the tools were woody twigs, and leaf-stalks or petioles accounted for almost another third. Put another way, only vines made up a similar proportion of tools at both places; all other raw materials' contributions were very different. More starkly, Gombe's chimpanzees *never* used leaves as

Table 7.5. *Lengths (cm) and diameters (mm) of termite-fishing tools at Gombe and Assirik*

Length	Gombe	Assirik
Number of tools	145	173
Mean	30.7	32.5
Median	28	30
Range	7–100	13–71
Diameter		
Number of tools	32	12
Mean	4	3
Median	5	3
Range	1–8	2–3

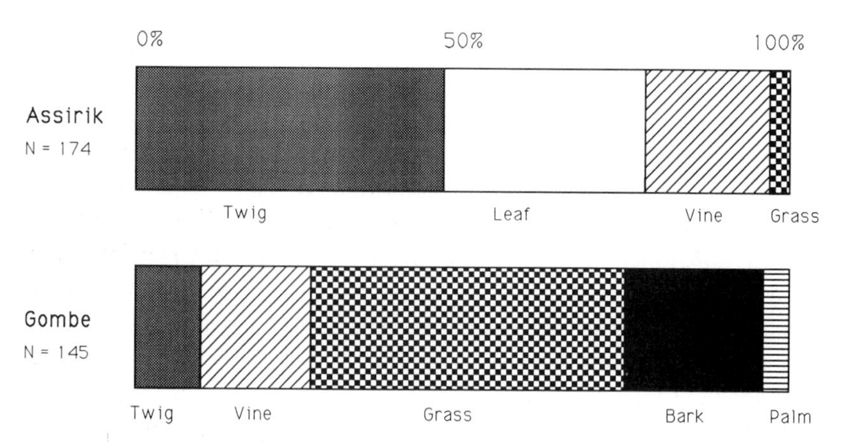

Figure 7.4. Types of raw materials used to make tools for termite-fishing, Assirik, Senegal, versus Gombe, Tanzania.

tools, nor did Assirik's *ever* use bark or palm fronds (see Figure 7.4). How can these differences in raw material be explained?

Both Gombe's and Assirik's apes occupy habitats with a wide variety of shrubs and trees (including palms) that provide twigs, bark and leaves, as well as suitable herbaceous and woody vines. A useful fishing probe must be both flexible and resilient, that is, it must be capable of bending slightly to conform to the twists and turns of the termites' passages, yet spring back upon withdrawal to its original shape for further insertions, if it is to be re-used (Teleki, 1974). Any of the above-named types of raw material will fulfil this function, given the right dimensions and proper processing. The nature of the task, defined by the structure of the mound, constrains the range of apt materials, but what is the basis for choice among them? Might this be a cultural matter?

One possible explanation is that the micro-habitats of termite mounds differ from place to place, and because chimpanzees take most of their tools from close by the site of use (see below), they may merely be passively taking whatever raw materials are there. Only one study has looked at such issues. At Assirik, McBeath & McGrew (1982) collected 323 tools in 25 assemblages from 15 mounds. They noted all plants (as potential sources of tools) growing within a 5-metre radius of 40 mounds distributed over five types of habitat. Over the termite-fishing season they checked 279 *Macrotermes* mounds for tools, and found that both tools and assemblages were greatly over-represented at mounds in the transition between open woodland and short-grass plateaux. Why is this?

The dimensions and density of mounds did not differ across types of habitat. Nor were chimpanzees concentrating their activities in the transition zone for other reasons. Nor did the transition zone offer the greatest overall abundance of raw materials or even the highest relative abundance of preferred types of raw materials. Only an analysis at the level of *species* of raw material yielded the answer: 80% of tools were made of the straight but limber woody shoots of *Grewia lasiodiscus* shrubs. These were concentrated in the transition zone. Thus, detailed analysis showed an environmentally deterministic explanation: chimpanzees' selection of raw materials for fishing probes mirrors the availability of the best raw materials, and any chimpanzee termite-fisher could come to such a sensible conclusion through individual trial-and-error, without need of tuition or imitation. There is thus no need to posit a cultural explanation.

However, the results of analyses of techniques of manufacture or modification are not so easily dismissed. Alterations to the raw material are usually done by reduction such as stripping of leaves, breaking off of twigs, peeling of bark, clipping of ends, etc., and this contrasts intriguingly between Gombe and Assirik. For example, woody vegetation can be partly or completely peeled of its covering bark. At Assirik, 86% of such twig or vine tools were totally peeled of their bark, which was always discarded (McGrew *et al.*, 1979*a*). At Gombe, *no* tool was ever peeled; instead in 21% of tools the bark was used for fishing (Figure 7.5), and the twig or vine was thrown away. Both populations knew how to peel bark, but they used the result in opposite ways! This sort of contrast looks by exclusion to be a social custom, a pattern 'liberated'

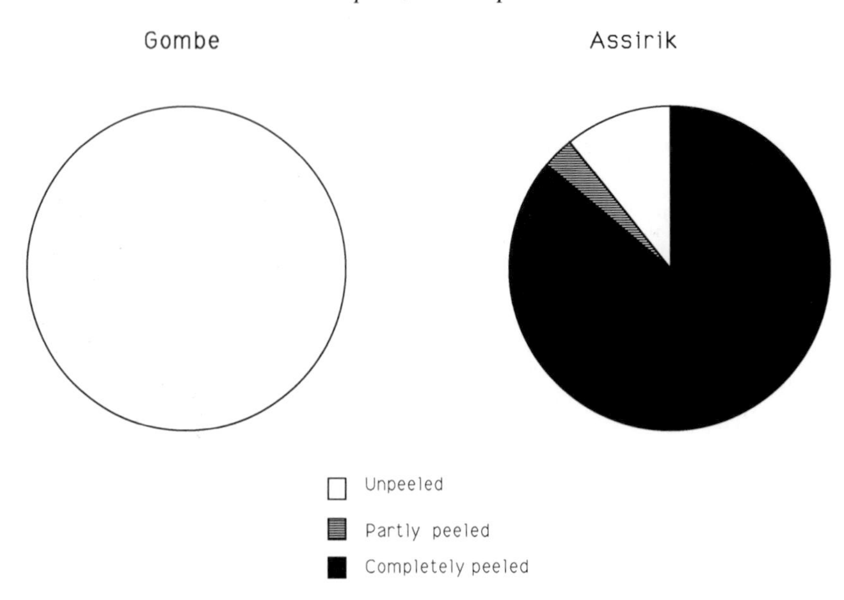

Figure 7.5. Peeling of bark from tools for termite-fishing, Gombe, Tanzania ($N = 36$), versus Assirik, Senegal ($N = 72$.) Data are percentages.

from environmental constraints. Not necessarily so. It is always possible (however unlikely) that Gombe's woody vegetation does not peel so well as Assirik's, or perhaps that *Grewia lasiodiscus* peels readily and immaculately, etc.

More convincing would be a difference between the two sites in the use of finished tools which is not a function of the raw material. Such a result emerged serendipitously. During a bout of fishing the chimpanzee inserted either one of the two ends of the tool into the mound. At Gombe this was directly seen, and at Assirik it was inferred from signs of wear and mud on the end(s) of the tool. Figure 7.6 shows that Gombe's chimpanzees usually used both ends of a tool, but Assirik's almost never did so (McGrew *et al.*, 1979*a*). Seeking to explain this contrast in terms of constraints imposed by the raw materials has an air of grasping at straws. For example, because vines are roughly uniformly cylindrical throughout their length, they are more likely to be used at both ends. But both populations used about equal proportions of vines (Figure 7.4), and the contrast is clear: *all* vine tools at Gombe were used at both ends and *all* vine tools at Assirik were used at one end only.

Observations of termite-fishing at Gombe (McGrew, unpublished data) suggested that the use of both ends of the tool by chimpanzee

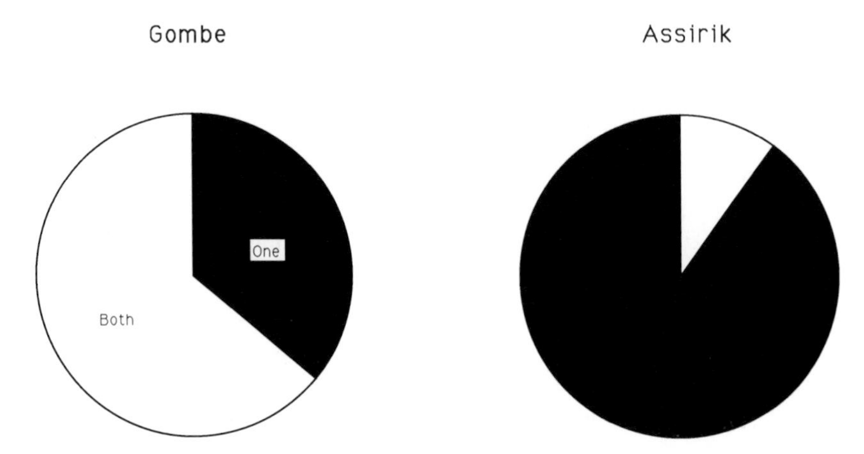

Figure 7.6. Use of one or both ends of tools for termite-fishing, Gombe, Tanzania (N = 53), versus Assirik, Senegal (N = 39). Data are percentages.

termite-fishers was non-functional. Instead, it looked like a 'superstitious' response to flagging returns from the last few insertions. Changing ends of the tools did not improve the 'catch', as the end of the tool used was irrelevant, just as changing hooks is of little use if the fish are not biting. Such persistent but useless habits seem just as likely candidates for being copied when a youngster learns to fish for termites by watching others as do useful ones (McGrew, 1977).

In conclusion, when one focusses empirically and comparatively on a particular kind of subsistence activity, termite-fishing, some contrasting features are less interesting, in that these are straightforward reflections of physical or biotic environment. By analogy, if one angler chooses to fish a river and another a lake, this may merely signify that the resources available differ. However, other features seem inexplicable in terms of such environmental transactions (to use Galef's, 1976, term) and are more likely to be understood in terms of the social milieu. Again by analogy, if one angler habitually threads a worm length-wise on the hook, while another loops a worm cross-wise, the explanation is likely (though not certain!) to lie in who taught them to fish, but not in the worm, the fish, the pond, etc.

Hammers and anvils

Stone tools have always been of special (and disproportionate?) significance to those interested in the evolutionary origins of technology (see

Chapter 9). Lithic artefacts preserve well, and so the earliest known tools in the palaeo-archaeological record have been interpreted as pebble hammers (Leakey, 1966; cf. Toth, 1985*b*). Given this, what can be said about non-human and especially chimpanzee use of hammers and anvils, across the board?

These apparently simple objects are vexingly hard to define with logical consistency. Hammers (*sensu stricto*) are here considered to be tools that are used in the hand to apply explosive percussive force to a resting goal-object. The result is that the goal-object is fractured to reveal its contents. Anvils (*sensu stricto*) are here defined as fixed objects, usually in the substrate, which support goal-objects to be hammered. Thus, anvils are not tools because they are not handled and remain stationary. In reality, goal-objects may be struck directly against a hammer, an anvil, one another, or the substrate. Also, anvils may be moved, although between and not during blows.

First, neither hammers nor anvils are unique to apes, or even to mammals. Song thrushes smash snails against stones embedded in the ground in order to crack open the molluscs so as to expose their edible body-parts inside (Henty, 1986). Conversely, Egyptian vultures use hammer-stones without anvils to crack open the eggs of ground-nesting birds (Goodall & van Lawick, 1966). California sea otters use portable anvils: they float on their back, balancing a stone on their chest, and smash molluscs against these anvils until they crack (Hall & Schaller, 1964). However, there seem to be no records of any non-primate using *both* hammer and anvil together.

Second, apart from chimpanzees, the only primates to use hammers *or* anvils habitually and spontaneously are capuchin monkeys. This occurs both in nature (Izawa & Mizuno, 1977) and in captivity (Visalberghi, 1987; Anderson, 1990), but there is a notable difference between these two conditions. All object-smashing confirmed for wild capuchins is anvil-use, while captive capuchins show anvil-use, hammer-use and hammer-and-anvil-use (Visalberghi, 1990).

Wild chimpanzees have been reported to use hammers and anvils at seven main sites: Bossou, Cape Palmas, Mt. Kanton, 'Liberia', Sapo, Tai and Tiwai Island. (Kortlandt (1986) reported a number of other subsites.) Free-ranging chimpanzees after release have been seen to do so at a further three places: Abuko, Assirik and Bassa Islands (see Figure 7.7; Table 8.1 gives details of references). However, behavioural data on the tools being used are available from only two of these: Tai and Bassa (Boesch & Boesch, 1983, 1984*a*,*b*; Hannah & McGrew, 1987).

Figure 7.7. Two-handed use of stone hammer and anvil to crack oil palm nuts, Bassa Islands, Liberia. (Courtesy of A. Hannah.)

Other reports are anecdotal, second-hand or circumstantial, and this may lead to confusion unless criteria for inference from indirect data are explicitly stated. (For examples of such standards applied to other tools, see McGrew *et al.* (1979*a* and McGrew & Rogers (1983).)

An example illustrates this problem. A 4-year study of the wild chimpanzees on the west side of Assirik recorded behavioural data on their smashing of hard-shelled fruits such as baobab against stone or root anvils and tree-trunks (Baldwin, 1979). This behavioural pattern is well known in other populations, both for hard-shelled fruits (Goodall, 1968) and for the crania of mammalian prey (Boesch & Boesch, 1989). A similarly long study of wild-born chimpanzees being rehabilitated on the east side of Assirik showed the same behavioural patterns (Brewer, 1978). However, while the rehabilitated chimpanzees also used stone hammers to open other hard-shelled fruits (e.g. *Afzelia africana*, *Oncoba spinosa*), there was no evidence for this in their wild counterparts.

Later, Bermejo *et al.* (1989) reported that the wild chimpanzees used hammers and anvils to smash open baobab fruits. All of the data were circumstantial, no criteria for inference were given, and the two photographs published could have shown either anvil *or* hammer use. Clearly, the phenomenon needs further study, perhaps by an inter-disciplinary team of primatologist and archaeologist.

In summary, chimpanzees are the only non-human species that spontaneously uses the hammer-and-anvil combination in nature. However, given all the precursory and variant patterns of behaviour now known, the data are sparse and need to be augmented if useful comparisons are to be drawn.

8
Chimpanzee ethnology

Cataloguing tool-kits

By the early 1970s it was clear that wild chimpanzees at various sites in Africa had different repertoires of tool-use. By then the *negative* evidence from Gombe's well-known subjects could be set against the scrappy positive evidence from elsewhere. Put another way, if the *tool-kit* of a group is defined as its complete set of tools and their use, then nothing has been added at Gombe since the early publications of Goodall (1964, 1968, 1973). Thus, from 1973 it could be said confidently that Gombe's apes did *not* use hammer-stones, and so a real difference existed between them and the chimpanzees of Cape Palmas, who were the subjects of the first anecdotal report of tool-use almost 150 years ago (Savage & Wyman, 1844).

Goodall (1973) produced the first catalogue of tool-use by free-ranging chimpanzees; it included 10 sites, and all but Gombe's data were based on short-term studies or single sightings (see Figure 8.1). Teleki (1974) followed with a list of 12 sites, of which only five were common to Goodall's catalogue of a year earlier! The most extensive published catalogue is that of Beck (1980), who compiled findings from 20 sites across Africa. More recently, Goodall (1986) produced another list, but this had only 16 populations of wild (but not released and free-ranging) chimpanzees. All of the previous efforts are now dated.

Table 8.1 lists 32 populations or groups of free-ranging chimpanzees in Africa that have shown some kind of tool-use, if a tool is defined as a 'moveable, inanimate object used to facilitate acquisition of a goal' (McGrew *et al.*, 1975). This minimal measure shows that the far western subspecies has 14 records, the central-western has 12, and the eastern has only 6. However, this apparent contrast is simplistic (see below).

No previous catalogue of tool-kits has sought to distinguish between

Study	No. of Populations
Goodall (1973)	**********
Teleki (1974)	************
Beck (1980)	********************
Goodall (1986)	****************
This study	*********************************

Figure 8.1. Catalogues of tool-use by free-ranging chimpanzees across Africa. Each asterisk represents a site where tool-use has been seen.

Table 8.1. *African study-sites of free-ranging chimpanzees at which tool-use has been recorded*

	Country	Subspecies[b]	Major sources
Abuko (r)[a]	The Gambia	v	Brewer (1978), Goodall (1973)
Assirik	Senegal	v	Baldwin (1979), Bermejo *et al.* (1989), McBeath & McGrew (1982), McGrew *et al.* (1979*a*)
Assirik (r)	Senegal	v	Brewer (1978, 1982)
Ayamiken	Equatorial Guinea	t	Jones & Sabater Pí (1969, 1971)
Baboon (r)	The Gambia	v	Brewer & McGrew (1990)
Banco	Ivory Coast	v	Hladik & Viroben (1974)
Bassa (r)	Liberia	v	Hannah & McGrew (1987)
Belinga	Gabon	t	McGrew & Rogers (1983)
Bossou	Guinea	v	Albrecht & Dunnett (1971); Sugiyama (1981, 1989), Sugiyama & Koman (1979, 1987)
Budongo	Uganda	s	Sugiyama (1969)
'Cameroon'	Cameroon	t	Merfield & Miller (1956)
Campo	Cameroon	t	Sugiyama (1985)
Cape Palmas	Liberia/Ivory Coast	v	Savage & Wyman (1844)
Dipikar	Equatorial Guinea	t	Jones & Sabater Pí (1969, 1971)
Filabanga	Tanzania	s	Itani & Suzuki (1967)

Table 8.1. *Cont.*

	Country	Subspecies[b]	Major sources
Gombe	Tanzania	s	Goodall (1964, 1968, 1970, 1973, 1986), McGrew (1974, 1977, 1979), Teleki (1974)
Ipassa (r)	Gabon	t	Hladik (1973)
Kanka Sili	Guinea	v	Albrecht & Dunnett (1971)
Kanton	Liberia	v	Kortlandt & Holzhaus (1987)
Kasakati	Tanzania	s	Izawa & Itani (1966); Suzuki (1966)
Kasoje	Tanzania	s	McGrew & Collins (1985), Nishida (1977, 1980*b*), Nishida & Hiraiwa (1982), Nishida & Uehara (1980), Uehara (1982)
Kibale	Uganda	s	Ghiglieri (1984, 1988); R. W. Wrangham (personal communication)
'Liberia'	Liberia	v	Beatty (1951)
Lopé	Gabon	t	Tutin *et al.* (1991)
Mbomo	Congo	t	Fay & Carroll (1992)
Ndakan	Central African Republic	t	Fay & Carroll (1992)
Ngoubunga	Central African Republic	t	Fay & Carroll (1992)
Okorobikó	Equatorial Guinea	t	Jones & Sabater Pí (1969, 1971), Sabater Pí (1974)
Sapo	Liberia		Anderson *et al.* (1983)
Tai	Ivory Coast	v	Boesch & Boesch (1990), Rahm (1971), Struhsaker & Hunkeler (1971)
Tiwai	Sierra Leone	v	Whitesides (1985)
West Cameroon	Cameroon	t	Struhsaker & Hunkeler (1971)

[a] (r), released populations.
[b] s, eastern subspecies; t, central-western subspecies; v, far western subspecies.

habitual versus *rare*, *idiosyncratic* or *questionable* tool-use by chimpanzees. (However, Sugiyama (1989) did distinguish between 'established' and other types at Bossou.) Thus habitual use is here restricted to patterns shown repeatedly by several members of a group. It excludes a single instance by one individual (Plooij, 1978*b*), a single instance by several individuals (Beatty, 1951), several instances by only one individual (Goodall, 1968), and all instances of insufficient data

Table 8.2. *Habitual patterns of tool-use of wild chimpanzees*

	Site							
Pattern	Gombe	Bossou	Kasoje	Tai	Kanka Sili	Assirik	Kanton, Sapo, Tiwai	Campo, Okorobikó	Kibale
Termite-fish	X		X			X			
Ant-dip	X	X		X		X			
Honey-dip	X			X					
Leaf-sponge	X	X							
Leaf-napkin	X								X
Stick-flail	X	X	X		X				
Stick-club	X	?X	X		X				
Missile-throw	X	X	X		X				
Self-tickle	X								
Play-start	X		X						
Leaf-groom	X		X						
Ant-fish			X						
Leaf-clip		X	X						
Gum-gouge		X							
Nut-hammer		X		X			XXX		
Marrow-pick				X					
Bee-probe				X					
Branch-haul		X							
Termite-dig								XX	
Total	11	8	8	5	3	2	(3×)1	(2×)1	1

X = present.

(Bermejo *et al.*, 1989). Cases by released chimpanzees ($N = 5$ in Table 8.1) are also omitted, because they may have been influenced or even shaped by their human caretakers. This is *not* to say that single cases are useless. On the contrary, it takes only one example to show a capacity (Brewer & McGrew, 1990). Many of the instances classed as non-habitual may merely be patterns awaiting more evidence. Finally, many of the older records are anecdotal because such natural history 'notes' were adequate by then-current standards of scientific reporting.

Table 8.2 presents a stricter catalogue, limited to habitual tool-use shown only by wild chimpanzees. Only 12 populations showed a total of 43 habitual tool-use patterns that met the above criteria, and within these the range was from one to 11 per site. Six of the sites showed but one pattern (Campo, Kanton, Kibale, Okorobikó, Sapo, Tiwai), but like Assirik ($N = 2$) all were places where the subjects of study were totally unhabituated or only minimally tolerant of close-range observation. Overall, a clear positive correlation emerges between degree of habituation or length of study and number of identified patterns of habitual tool-use, which suggests that the results are incomplete for most populations.

No pattern of tool-use even comes close to being universal. The most widespread, the use of a hammer to crack open nuts, is known at only five sites. The inter-related agonistic or anti-predatory patterns of weapon-use (flail–club–missile) occur in four populations, but one of these, Kanka Sili, was experimentally induced (Albrecht & Dunnett, 1971). More impressive are the four occurrences of ant-dipping, especially as this pattern was found from the wettest forest site at Bossou to the driest savanna site at Assirik.

The *functional* nature of the habitual tool-use is instructive: most (21 of 43) are subsistence activities of acquiring or processing food, especially social insects ($N = 13$) or nuts ($N = 5$) (see Figure 8.2). Only one each relates to meat, water, or other plant foods. As noted above, the weapon-use total ($N = 12$) may be inflated by its make-up of related patterns. The 10 remaining habitual types are split between self-directed (leaf-napkin, self-tickle) and apparently ritualised communicative signals (play-start, leaf-groom, leaf-clip).

Non-subsistence technology

Single studies and comparative analyses of tool-use by free-ranging chimpanzees have concentrated overwhelmingly on subsistence activi-

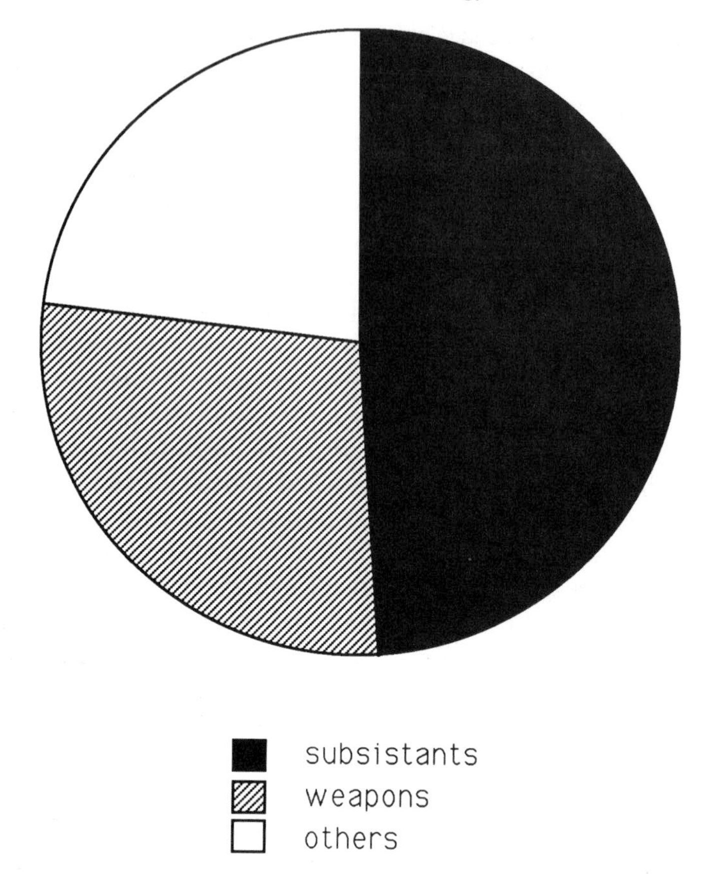

Figure 8.2. Functions of tools used habitually by chimpanzees over all wild populations studied so far. Data are percentages.

ties, that is, on the finding, capturing and processing of energy and nutrients. This is entirely appropriate, as most chimpanzee tools are subsistants (see Figure 8.2), but other aspects of technology may be more revealing, at least in principle.

Best documented is perhaps the most marginal type, that of the medicinal consumption of vegetative plant parts by wild chimpanzees. This is not tool-use in the normal sense, but it is not usual subsistence either: leaves, bark or pith are processed for their bio-active ingredients, and the 'containers' are discarded or passed virtually unaltered through the gut. For chimpanzees, Wrangham & Nishida (1983) first recorded such self-administration of *Aspilia* leaves at both Gombe and Kasoje. Later biochemical analysis revealed these to contain a potent antibiotic, hiarubine A, with anti-fungal, -viral, -bacterial and -parasitic properties (Rodriguez *et al.*, 1985). Later, at Kasoje, other plants have been seen to be eaten similarly, suggesting other chemical functions (Takasaki & Hunt, 1987; Huffman & Seifu, 1989). In the latter report a case study of relief of symptoms in an adult female chimpanzee was recorded. All species of plants used by the apes are also used medicinally by local human populations.

Otherwise, most non-subsistence tool-use is poorly known. First, there are revealing anecdotes such as Sugiyama's (1969) two instances of chimpanzees at Budongo using leafy-twigs to shoo away flies. This shows the 'minimal necessary competence' (to use Wynn's, 1989, term) to make a simple fly-whisk, but little more. The repeated use of empty paraffin tins by a challenging adult male to enhance his agonistic display was similarly fascinating but idiosyncratic (Goodall, 1971). The tins were artificially introduced and then removed when their disruptive potential was realised. Further, there are cases when more than one chimpanzee repeatedly shows tool-use with several variants, *but only within a very specific context.* Captive adolescent chimpanzees at the Delta Primate Center performed dental grooming and extractions with tools of wood and cloth, but only during a period when they were shedding their milk teeth (McGrew & Tutin, 1972, 1973) (see Figure 8.3).

More frustrating are three kinds of chimpanzee tool-use that are often seen but virtually unstudied. From the early research of Köhler (1927) onwards, many researchers watching chimpanzees in zoos or laboratories have seen them use probes or prods to investigate the environment (Beck, 1980). Straws may be poked into cracks or sticks against novel objects. The recipients of probing or prodding may be animate or inanimate, either apes or other species, yet there seems to be no comprehensive descriptive account for *any* population, much less a systematic study.

Similarly, many observers of normal infant and juvenile chimpanzees know that they use a variety of objects, almost anything that they can get hold of, in both self-stimulation and social interaction (McGrew, 1977). Self-directed use of stones, sticks, leaves, food-items, etc., may be playful (as in self-tickling), sexual (as in masturbation) or exploratory (as in probing in nooks and crannies) (see Figure 8.4). Social use of such objects may also be serious practice for later life, as in agonistic flailing, clubbing or throwing, but more common is the use of objects to start play with peers (Plooij, 1978*a*) (see Figure 8.5). Teasing invitations over

Figure 8.3. Dental grooming, Delta Primate Center, USA: (*a*) dentist opens patient's mouth; (*b*) dentist examines patient's deciduous teeth; (*c*) dentist uses tool to probe teeth (right hand) while holding lower jaw (left hand); (*d*) spectator watches dentist use probe on patient.

(c)

(d)

Figure 8.3 *cont*.

Figure 8.4. Infant plays with palm frond while clinging to his mother, Gombe, Tanzania.

Figure 8.5. Two infants carry twigs in their mouths during social play, Gombe, Tanzania.

possession of an object (even if it has no intrinsic value) and 'catch-me-if-you-can' fleeing are daily occurrences when nursery groups of mothers and their young offspring form (Plooij, 1978*a*; Adang, 1986). For investigatory probing, self-tickling and play initiation, no comparative analyses can be done until findings are collected and presented.

Extensive data *are* available on weapon-use by the wild chimpanzees at Gombe (Goodall, 1986). Sticks and stones were flailed, clubbed, and thrown at other chimpanzees and at other species, chiefly olive baboons and humans. A fourth behavioural pattern, whipping with still-attached vegetation, is not therefore tool-use but is closely related in function. Also related is dragging, in which a log or branch is pulled along the ground behind a displaying individual (Boesch & Boesch, 1990). This is tool-use but is not weapon-use, as it is not directed at a target.

Weapon-use at Gombe was predominantly shown by males: over 6 years of pooled data, the frequency of males' flailing, clubbing and throwing was almost eight times that of females. Clubbing was too rare (only 6% of cases of weapon-use) for further analysis, but there seems to be a difference in targets between the other types: almost half of flailings were directed at other chimpanzees compared with fewer than a third of throws.

Spontaneous weapon-use by chimpanzees has long been known for both captive (Kortlandt & Kooij, 1963) and wild chimpanzees (Sugiyama, 1969; Nishida, 1970), but no systematic analyses have been published. Especially striking are the graphic descriptions and films of chimpanzees responding with induced weapon-use to the sudden presentation of a stuffed (and sometimes moving) leopard (Kortlandt, 1965; Albrecht & Dunnett, 1971). Unfortunately, no statistical analyses were given, so claims of differences between forest-living versus savanna-living chimpanzees in their reactions must remain as hypotheses yet to be tested. It seems remarkable that behavioural patterns that have played such an important part in evolutionary reconstructions (Kortlandt, 1980) should remain so empirically neglected. Goodall's studies (1986) are a welcome start, but comparative analyses await records from other sites, e.g. Kibale (R. Wrangham, personal communication).

Goodall (1986) also reported the only systematic analyses of tool-use in personal hygiene, again at Gombe. She reported 230 instances of leaves being used as napkins; in 90% of cases the substance removed was one of four types of bodily fluid: semen, faeces, blood or urine. After mating, males were more fastidious than females; they wiped their penes over 10 times more often than females wiped their vulvas. No such data are available from any other population, although the use of napkins occurs elsewhere (Tai: Boesch & Boesch, 1990; Kasoje: T. Nishida, personal communication; Kibale: R. Wrangham, personal communication).

Two other patterns involving leaves are perhaps the most esoteric of all chimpanzee tool-use: leaf-clipping and leaf-grooming. In leaf-clipping, the performing chimpanzee noisily pulls to bits one or more leaves by hand and mouth, leaving only the stripped petiole (Nishida, 1980*b*). The result most closely resembles a fishing tool, but the function is completely different, being most probably a signal. At Kasoje, 56% of cases of leaf-clipping were in courtship, usually directed by a male to an oestrous female. In most other instances the leaf-clipper was apparently frustrated, most often by lack of access to a tempting incentive such as food processed by others.

Thousands of kilometres away at Bossou, the same pattern functioned similarly but not identically (Sugiyama, 1981). There, only 7% of cases were in sexual contexts, but 48% were done in clear frustration, and a further 36% occurred in frustration-related aggression when the chimpanzees sought to drive away a persistent human observer. In almost all cases at both sites the leaf-clippers seemed to be in approach-avoidance conflict, so that the result looked like a ritualised displacement activity. No other population of chimpanzees, wild or captive, has been reported to do leaf-clipping.

Leaf-grooming is more enigmatic. Goodall (1968) first described this calm and deliberate custom at Gombe as occurring when a chimpanzee directed typical grooming motor patterns (such as manipulate, peer, mouth, lip-smack) to randomly picked leaves. It was not directly functional, in that the leaf was not cleaned. Plooij (1978*a*) classed leaf-grooming as 'proto-declarative', that is, the use of an object to get another's attention. Wrangham's (1980*b*) detailed analysis showed it to be linked to true grooming, where it served to start or to perk up flagging grooming bouts with others. More rarely it occurred when a lone chimpanzee seemed bored, and Goodall (1986) likened this solitary leaf-grooming to doodling. The pattern was seen daily at Kasoje (Nishida, 1980*b*), and also occurs at Kibale (R. Wrangham, personal communication) but no analysis has yet been presented.

In summary, the potential for comparison across populations of non-subsistence tool-use is disappointing, apart from weapon-use. For weapons, the behavioural constellation of related patterns is known to be widespread, but the data so far presented are not even enough to tackle the most basic issues, such as whether or not the male predominance seen at Gombe is a universal one. Most of the other patterns depend on thoroughly habituated subjects to allow collection of enough data, either because the mothers of young infants are notoriously shy (for self-tickle, play initiation) or because the patterns are inconspicuous or rare (leaf-clip, leaf-groom).

Regional and local patterns

Regional and local variation are taken for granted in ethnology and palaeo-anthropology when the subjects of study are human or near-human beings. Thus, Ucko (1970) found differences in penis sheaths across South America, Africa, New Guinea and the Southwest Pacific, as well as within these regions. Similarly, Wynn & Tierson (1990) found differences in the shape of late Acheulean handaxes across Europe, Africa, India and the Near East. To examine the possibility of variation in space for chimpanzee tools requires a historical perspective.

If the 1960s can be characterised as the initial period of descriptive ethnography of chimpanzees, then the 1970s can be thought of as the decade of gross, regional comparisons, followed in the 1980s by finer-grained, local comparisons. As discussed above, until Goodall (1964) showed that Gombe's chimpanzees had a tool-kit (although she did not use the term), all previous accounts were one-off anecdotes. Such minimal ethnography continues to fill in gaps, especially in populations whose habituation is incomplete. For example, it is useful to note that ant-dipping at Bossou was first seen only in 1987 (Sugiyama *et al.*, 1988), though the chimpanzees had been studied since 1976.

The first attempt to compare different populations of chimpanzees was by Struhsaker & Hunkeler (1971) (see Table 8.3). They set a pattern of using the geographical races of subspecies of chimpanzees as a basis for regional comparison. They hypothesised that far western chimpanzees were hammer-users to smash nuts while eastern and central-western chimpanzees were termite-'fishers'. The latter entailed lumping all forms of tool-use to get termites. Nishida (1973) followed the same dichotomous distinction, but expanded fishing to include ants as well as termites, given his findings on *Camponotus*-eating at Kasoje.

Teleki (1974) refined this scheme into the first three-way one, whereby far western chimpanzees pounded with hammers, central-western chimpanzees probed for termites (but not ants), and eastern

Table 8.3. *Hypothesised regional differences in tool-use across wild chimpanzees*

Eastern (*P. t. schweinfurthii*)	Central-western (*P. t. troglodytes*)	Far western (*P. t. verus*)	Sources
Termite-'fishers'		Nut smashers	Struhsaker & Hunkeler (1971)
Ant/termite-fishers		Nut smashers	Nishida (1973)
Foliage-industry	Stick-industry	Stone-industry	Sabater Pí (1974)
Ant/termite-probers	Termite-probers	Pounders	Teleki (1974)
Termite-fishers	Termite-diggers	Termite-fishers	McGrew *et al.* (1979*a*)

chimpanzees probed for both ants and termites. This distinction was untenable even before it appeared, as Hladik (1973) had reported extensive ant-fishing at Ipassa. At the same time, Sabater Pí (1974) presented another three-way distinction, based on stones (far western), sticks (central-western) and foliage (eastern) (see Figure 8.6). Already published data belied this too, as Goodall's (1964, 1968) reports of ant-dipping with sticks were well-established. Both Teleki's and Sabater Pí's comparisons ignored Goodall's (1973) report of Brewer's observations of wild-born, rehabilitated chimpanzees in The Gambia probing for termites, thus making far western chimpanzees more than nut-crackers.

McGrew *et al.* (1979*a*) presented a new kind of regional comparison. First, they held prey constant (*Macrotermes* termites) and concentrated on one kind of tool-use (probes of vegetation). Second, they presented detailed and systematic comparisons of features of tool-use, such as dimensions of tools. Third, they related differences in tool-use to differences in the ecology of prey and raw materials, such as patterns of rainfall and termites' mound-building. Finally, in addition to hypotheses, they gave explicit ways in which these could be falsified. The result was a scheme whereby forest-living (central-western) chimpanzees were termite-diggers and savanna-living (far western) and woodland-living (eastern) chimpanzees were termite-fishers. These hypotheses *were* falsified, like their predecessors, by McGrew & Rogers' (1983) report of termite-fishing in the forest at Belinga in the central-western region.

So, what is the current state of regional comparisons? Some differences still persist, as Tables 7.3 and 8.2 show. Use of hammers of stone or wood to crack open hard-shelled containers has still been

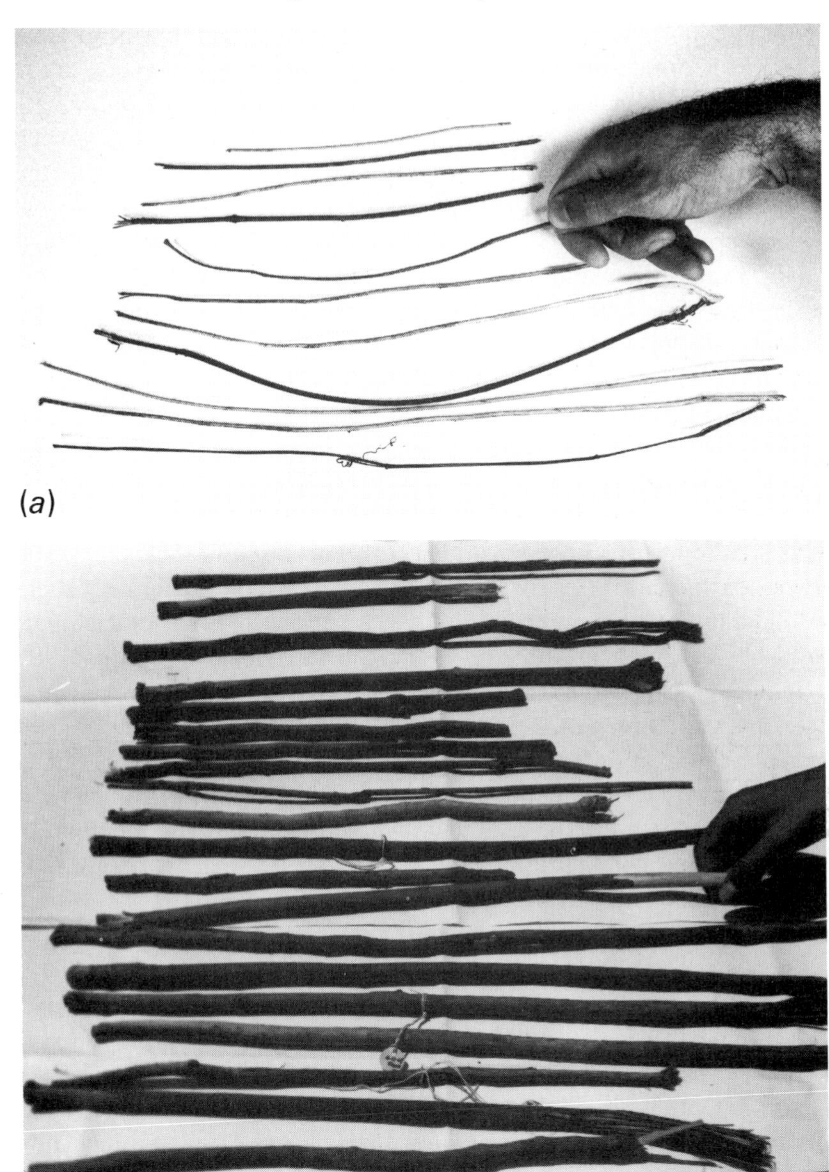

Figure 8.6. Tools used to get termites (and human hands): (*a*) fishing probes, Assirik, Senegal; (*b*) digging sticks, Okorobikó, Equatorial Guinea. (Courtesy of J. Sabater Pí.)

reported only in the far west beyond the Dahomey Gap. Use of sticks as picks to perforate or to dig up termites' mounds remains unique to the central-western chimpanzees of the equatorial forests, as apparently does the use of the brush-stick (Sugiyama, 1985). Use of medicinal plants in self-treatment of illness has been recorded so far only from the eastern subspecies of the chimpanzee (Huffman & Seifu, 1989). All other apparent regional distinctions have fallen by the wayside, though many await further study; for example, Sugiyama's (1989) recent report from Bossou in Guinea of leaf-sponging to get water out of tree-holes means that this behavioural pattern can no longer be thought of as unique to Gombe and Kasoje in Tanzania. Thus, like leaf-clipping at Bossou and Kasoje, it must now be watched for elsewhere.

The first attempts to contrast behaviour *within* regions were again those of Goodall (1973) and Nishida (1973). This led to explicit, point-by-point comparisons of diet (McGrew, 1983; Nishida *et al.*, 1983) and social customs (McGrew & Tutin, 1978). For tool-use, the first specific two-way comparison was of termite-fishing at Gombe and in the Mahale Mountains (Nishida & Uehara, 1980). Recent discoveries of such patterns as leaf-sponging and honey-dipping at Kibale in Uganda (R. Wrangham, personal communication) will allow further such exercises.

More importantly, this led to another type of more precise local comparison, that of the tool-use patterns in neighbouring communities or groups within a population. Uehara's (1982) preliminary findings and hypotheses about differences in ways of getting termites among B-, K- and M-groups' chimpanzees at Kasoje were tested extensively by Collins & McGrew (1985, 1987; McGrew & Collins, 1985).

Similarly, using Sugiyama's data from Bossou, Kortlandt (1986) proposed two types of hammer-use for far western chimpanzees. Type I was small stones used to crack palm nuts; Type II was larger stones and wooden clubs used to crack other nuts. Type I was hypothesised to be limited to Bossou, while Type II was found in humid evergreen forests across hundreds of kilometres of southern Sierra Leone, Liberia and Ivory Coast. Later work (Kortlandt & Holzhaus, 1987) falsified this dichotomy, with the discovery of Type II cracking of *Coula edulis* only 13 kilometres west of Bossou.

The overall picture of variety in chimpanzee tool-use has seen a pendular change. From a position of generalising about the species as recently as 25 years ago, the tendency now is to emphasise differences, not similarities (Boesch & Boesch, 1990; Sugiyama, 1992). Contrasts are well-known across regional races, populations, and communities or groups. However, establishing the existence of differences says nothing in itself about their origins, which is the subject of the next section.

Innovation

Contrasts in behaviour between groups need not be cultural. Differences can be innate, just as similarities can be learned, to use the old-fashioned nature–nurture terms. Neither does complexity nor variance in behavioural patterns tell us whether they are of cultural or idiosyncratic origin. The simplest pattern may be acquired by imitation, and phenotypic variety may reflect genotypic variation, as in pleiotropy.

The only way to be confident of the cultural nature (!) of an act is to see it being done for the first time by an individual and then passed on to others. Thus, innovation and dissemination are needed. These are the only first two of the eight conditions set out in Chapter 4, but all others follow from them.

There seem to be at least four ways that this could happen (cf. Kummer & Goodall, 1985). The clearest would be the spontaneous *invention* (in a constant environment) of a new pattern that was then copied by others. This could occur in an intact group and would be recognised by the novelty of the act set against a history of negative data. A second way is by *diffusion*, in which an already-skilled performer joins a naive group and shows them (inadvertently or not) the pattern. A third way is when the new pattern is prompted by *environmental change*. An existing habit may be altered by natural forces or the group may occupy a new habitat and its members adapt to the change. A fourth way is really a special case of the third, when the agent is *intervention*. Sympatric humans (or other species) may intentionally or accidentally shape the behaviour of the chimpanzees. Any of these four types of innovation, if followed by dissemination, could change the technology of a group of apes.

Taking these in reverse order, there is much evidence of the effects of human influence, especially as most behavioural data come from provisioned subjects. Chapter 4 cited the spread of the use of levers to open banana distribution boxes, and described the famous example of Mike and the paraffin tins, both at Gombe (Goodall, 1973). Russon & Galdikas (1992) listed an extraordinary range of tool-use activities acquired imitatively by rehabilitated orang-utans from their human caretakers: siphoning, weeding, bridge-making, sweeping, tooth-brushing, canoeing, etc.

More hypothetical, but equally inadvertent, is Kortlandt's (1986) view that the wild chimpanzees of Bossou learned to crack palm nuts from the local humans. He stated that the work-sites are identical in appearance, distinguishable only by their locations. (Whatever the similarities, equally plausible would be the opposite conclusion, that the humans originally learned the pattern from the apes.) Even more speculatively, Eaton (1978) claimed that chimpanzees must have learned to use weapons from watching early hominids using these aids to deter large carnivores. Less disputable are *deliberate* human interventions, such as Brewer's (1978, 1982) teaching of her rehabilitants to use stones as hammers to smash open *Afzelia* pods, a custom never seen in wild apes.

There seem to be no cases of natural environmental changes being seen to cause innovation in tool-use by chimpanzees. The closest case may be the use of hammers to crack open oil palm nuts, in that the wide range of feral oil palms is a recent development on an evolutionary time-scale. It is not known how the oil palm spread so widely, but it was probably a combination of natural seed dispersal and unnatural human horticulture. In any event, the availability of palm nuts does not ensure their use, with or without the help of tools, as Chapter 1 showed. Kasoje's chimpanzees present several well-recorded examples of dietary innovations, all of them cases of the apes making use of feral cultigens such as mangoes originally planted but then left behind by departed villagers (Takasaki, 1983; Takahata *et al.*, 1986). This was occupation by the apes of an empty but hardly natural 'niche'. The simplest explanation for the lack of innovation in response to environmental change is that the time-scale of field primatology is too short to pick up all but the most catastrophic of natural changes, none of which has yet been documented. (A prime candidate for this would be the effect of local forest clearance, however.)

The arrival of tool-use by immigration has probably occurred at Kasoje, in that termite-fishing females from K-group have moved to M-group (Takahata, 1982). Uptake of the pattern by the residents, that is, diffusion, remains to be seen. A more striking case was the dramatic spread of palm nut cracking on one of the Bassa Islands, as reported by Hannah & McGrew (1987). For 12 weeks before the arrival of an adult female to join the group of rehabilitants on the island, no signs of tool-use were seen. Within hours of arrival she began to use stones to crack nuts, and within a month nine of the 13 chimpanzees were also cracking nuts. Strictly speaking, this does not show social learning (cf. Whiten,

1989; Tomasello, 1990), as the other chimpanzees may already have known the pattern and only have been 'prompted' by her enthusiastic performance to recall it. However, this seems far-fetched.

Finally, there is no known case of spontaneous invention of a tool-use pattern, followed by dissemination, in the pure sense (cf. Kummer & Goodall, 1985). That is, all known cases can be alternatively explained by the previous three ways. Perhaps this should not be too disappointing. How many studies, however long term, of foraging peoples *in situ* have ever reported spontaneous invention of new tools? Or, if the Acheulean handaxe persisted virtually unchanged for hundreds of thousands of years, why should we expect to be lucky enough to see notable changes in the material culture of apes in three decades? In the absence of data, one can turn to sheer speculation, in the time-honoured tradition of prehistorians. Hence the next section!

The invention of termite-fishing

Like many other animals, from birds to humans, chimpanzees gorge themselves on the winged reproductive forms (alates) that are released from termite mounds in the rainy season. These slow-flying, energy-rich prey exit from a few, slot-like holes and can be picked off easily as they emerge. Also concentrated at these exit-holes are soldier termites (to repel predators) and worker termites (to open and close the holes), for which this is a rare above-ground exposure. (Both castes are totally blind, unlike the alates.)

On an unrecorded but momentous occasion when there were many non-reproductives on the surface, an innovative Chimpanzee accidentally picked up a soldier along with an alate. It tasted good too. However, the next time she tried to pick up a soldier, it bit her on the forefinger with its powerful mandibles, drawing a tiny drop of blood. Curious and fearful, she picked up a nearby twig from the ground and poked at the insects on the ground. (This use of an investigatory probe is a common chimpanzee behavioural pattern.) Reflexively the termites attacked the prodding object, affixing themselves to it with their pincers. After peering at the insects, the chimpanzee nibbled off the prey from the twig, much like nibbling berries from a branch.

Meanwhile the flight of alates ended, and the soldiers retreated underground, leaving only a few to guard the exit-hole while the workers re-sealed it. The ape's repeated poking was thus 'led' to the hole, then into it, in pursuit of the vanishing prey. Deeper and deeper insertions into the mound were needed to keep the bout going, and these often bent or damaged the probe. The tool thus had to be replaced, and when convenient stray bits of vegetation were exhausted, the chimpanzee pulled off whatever living-parts of suitable shape and size were growing close at hand.

Eventually, the hole was exhausted and the insects stopped biting, but when the ape turned to the other holes that had been releasing alates, these had by then been sealed. However, the outline of the hole remained barely visible in the freshly-worked clay, and a flick of finger or thumb-nail re-opened it. The fresh odour of termite pheromones wafting from the hole reinforced further exploration, and an inserted probe yielded more termites. (These characteristic exit-holes remain visible until the next heavy rain washes them away, so that the stimuli will prompt a chimpanzee's memory even when no alates are being released.) *Termite-fishing had been invented.*

How likely is this scenario of invention, as brain-stormed over breakfast at Gombe one morning by Caroline Tutin, John Crocker and me? It is seductively plausible, but other alternatives, such as generalisation of motor patterns from ant- or honey-dipping, are also possible. We are unlikely ever to know what happened historically. Termite fishing may just as well have been invented in 1959, the year before Jane Goodall arrived, or a million years ago. It may well have been invented only once by some chimpanzee Edison or Logie Baird, and then diffused across Africa, or it may have been re-invented 100 or 1000 times. Until more sophisticated analytical techniques are devised, these remain tantalising, open questions. However, in a neat bit of reverse modelling, Anthony Collins (unpublished data) has shown that the whole sequence can be functionally acquired by a naive human primate.

Cross-cultural chimpanzees?

Does the above material establish a case for the material culture of chimpanzees? Is there now scope for a Chimpanzee Relations Area File (Murdock & Provost, 1973), or a CCCCCC (Chimpanzee Cross-Cultural Cumulative Coding Center)? The answers to these questions are not straightforward, and hark back to many of the issues raised in Chapters 1 and 4. Two points are worth emphasising. First, there seems little doubt that the wealth of data on chimpanzees exceeds the capacity of the traditional descriptive techniques of natural history. Were A.H. L.F. Pitt-Rivers alive today he might note an uncanny replication of the cumulative ethnography of the nineteenth century. For methodological reasons alone, the more powerful analyses of modern ethnology need to be applied. Second, if the contents of this chapter were reported unchanged except for a single independent variable – species – then the answer to the question opening this paragraph would be taken for granted as positive. If the same data were reported in ethnological journals as cross-cultural comparisons of human beings, not an eyebrow would be raised. (This is exactly the position for the ethno-botany of medicinal use of wild plants recounted earlier.) To paraphrase Louis Leakey, we must change our definition either of humanity or of culture, for we can no longer have both.

9

Chimpanzees as models

Kinds of models

Knowledge of chimpanzees has been explicitly built into models of human evolution for at least 30 years, since the emergence of sometimes startling findings from modern field-studies (Kortlandt & Kooij, 1963; Goodall & Hamburg, 1974; McGrew, 1979, 1981*a*; Tanner, 1981, 1987; Ghiglieri, 1987; Wrangham, 1987). Other species of African primates have also been cited in reconstructions of hominisation: savanna baboon (Washburn & DeVore, 1961), gelada (Jolly, 1970), bonobo (Zihlman *et al.*, 1978). Such models abound. Foley & Lee (1989) listed nine published between 1963 and 1987. This modelling has ranged from speculative outlines (scenarios, just-so stories, evolutionarios?) to systematic, point-by-point formulations (Wrangham, 1987; Wynn & McGrew, 1989). Further, seemingly countless numbers of articles on primate and especially chimpanzee natural history have ended with an apparently obligatory final paragraph on the implications for human evolution. How can we make sense of and choose between these many options?

One starting point is to distinguish between two main types of model: *referential* (Tooby & DeVore, 1987) or *analogous* (Dunbar, 1989) versus *conceptual* or more specifically *strategic* (Tooby & DeVore, 1987). The former make use of a known phenomenon such as the living chimpanzee as a referent for an unknown phenomenon such as an extinct proto-hominid. The latter use basic evolutionary and ecological theory as developed from studies of all living organisms to construct a tailored set of principles to elucidate the absent proto-hominid. Each type of model has its advantages and disadvantages (cf. Tooby & DeVore, 1987), but when used thoughtfully both yield testable hypotheses, in the form of predictions or post-dictions to explain the data.

(Recently, Stanford & Allen (1991) have argued that the two types are indistinguishable when scrutinised, that is, all conceptual models must at some level be based on living primates.)

Until recently, referential models held sway, as debate focussed on which living species was reckoned to be the most useful. The leading candidates were usually the chimpanzee on phylogenetic grounds versus the baboon on ecological grounds (Dunbar, 1989). Thus, the ape was advocated on grounds of *homology*, that is, on the basis of phylogenetic descent, while the monkey was argued for by *analogy*, that is, on the basis of convergence in only distantly related forms. In the later 1980s conceptual models gained in standing, as the implications of the current synthesis of evolutionary ethology and ecology (or sociobiology) filtered through (Steele, 1989). Yet this volume uses a referential model, based on the chimpanzee, and goes against the current grain, so it must be justified.

The main advantage of a referential model is its concreteness, especially from the viewpoint of empirical testing. For living chimpanzees and human beings there are both existing and potential quantitative data. These can be used precisely to compare and to contrast, with the goal of pin-pointing the absent, unavailable milestones in hominisation. Chimpanzees and modern humans can be thought of as the respective endpoints that define gaps ranging from narrow or non-existent (visual acuity), through bridgeable (food-getting), to wide (written language). These similarities and differences suggest ways of filling the gaps, that is, they point to data to re-examine or to collect anew. To give a specific example, prey size seems to be the variable that limits hunting of mammals by humans and apes. This spurs us to look at why chimpanzees take neither very small (< 1 kilogram, such as rodents) nor large (> 15 kilograms, such as adult ungulates) prey, while humans do, and to focus on the intermediate range of prey of about 5 kilograms where both referential forms overlap.

No existing conceptual or strategic model is this precise. Most are qualitative and descriptive. The most extensive such model, that of Tooby & DeVore (1987), is admirably comprehensive and erudite, but it makes no specific, testable predictions. More recent attempts do (e.g. see Foley & Lee's, 1989, social states and evolutionary pathways, or Dunbar's, 1989, top-down analyses), but it seems likely that referential models, despite their limitations, will go on serving usefully in the foreseeable future, along with conceptual ones.

Put another way, we can do empirical science directly on a present organism while we can only guess about an absent one. Chimpanzees referentially provide *both* tools and the *acts* of their making and use. No other referential model nor any conceptual model meets these simple conditions.

Models of what?

If chimpanzees are proposed to be useful models, the obvious next question is 'Of what?'. Until recently, evolutionary models, both referential and conceptual, were either imprecise or unconvincing. Most referred to 'early hominids' (Lovejoy, 1981) or 'proto-hominids' (Isaac, 1978*b*), or if a specific taxon was mentioned, such as *Australopithecus* by Tanner (1981), the rationale for its choice was not made explicit.

Over the last decade the stages of hominisation targeted in modelling have increased in number and specificity. The late Glynn Isaac (1978*a*) made chimpanzees the starting point for a three-way comparison, the other two being Plio-Pleistocene proto-hominids and living hunter-gatherers. Tanner (1981, p. 19) focussed on three critical stages: pre-hominid ancestral, transitional, and earliest hominid (*Australopithecus*). More recently, Tooby & DeVore (1987) gave eight reference points: ancestral hominoid, early australopithecine, later australopithecine, transitional form to *Homo erectus*, *H. erectus*, archaic *H. sapiens*, Neanderthal, anatomically modern human. This seems excessive, unless a detailed model is produced that justifies such splitting. Foley & Lee (1989) sought to apply their modelling to four stages (ancestral hominoid, first hominid, early *Homo*, modern human), and what follows below is close to their scheme (cf. McGrew, 1989*c*). My aim here is to focus on the evolution of material culture and to try to find the stage that gives the 'best fit' to the chimpanzee.

The safest choice on grounds of homology is a Miocene hominoid descended from some small, gibbon-like, stem ape. Thus, one seeks the last common ancestor (*Proconsul*?) before the split of the pongid and hominid lines in Africa, sometime between 4 and 7 million years ago (ma). The immediate problem is that there are no known artefacts from that period, and such an African ancestral ape may have been an accomplished tool-user like a chimpanzee or a non-tool-user like a gorilla or something in between (McGrew, 1989*c*). Also, there are no pongid fossils from that period and few hominid ones, to provide anatomical clues (Foley, 1987*a*). So, there are no artefacts yet found to model, but at least knowledge of tool-use by living chimpanzees can be used as a guide to know what to look for.

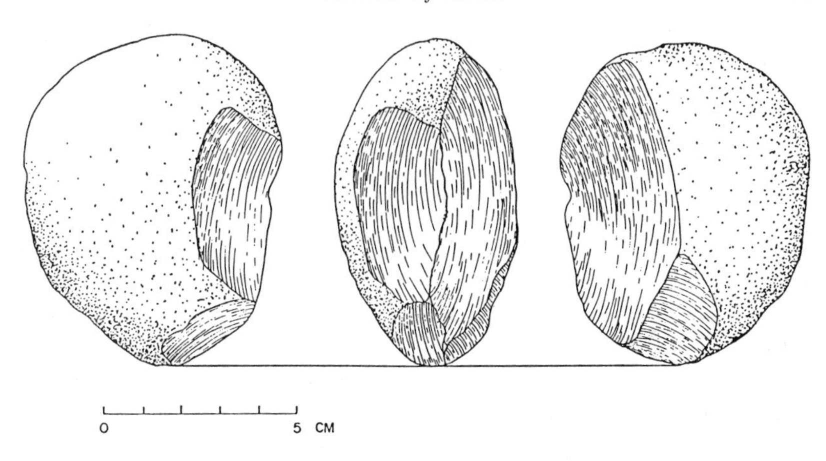

Figure 9.1. Oldowan stone tool. (Courtesy of T. Wynn.)

The next choice on some mixture of grounds of homology and analogy is a Pliocene proto-hominid, an australopithecine of some 3–4 ma. By then there was significant structural change in hominid evolution (e.g. emergence of bipedalism) but no real 'explosion' in brainpower, as well as at least two confusing taxonomic radiations in eastern and southern Africa. The appearance of undeniable artefacts in the palaeo-archaeological record came later, such as the crude stone tools from Hadar dated at 2.4–2.7 ma (Harris, 1983) (see Figure 9.1). Pertinent questions for investigation may be whether or not living apes could and would make such tools and, if so, for what?

The first choice on grounds of clear analogy would be the earliest known Plio-Pleistocene large-brained hominid (*Homo habilis*?) as lived at Olduvai Gorge, Tanzania, or East Turkana, Kenya, some 1.5–2.0 ma. This Oldowan lithic culture differs from anything so far known for free-ranging apes, but all of the equivalent properties are found in the non-lithic tools of chimpanzees (Wynn & McGrew, 1989). Complicating the picture are persisting smaller-brained hominids in the form of robust australopithecines, who also could have made the tools (Brain *et al.*, 1988; Foley, 1987*b*). The challenge is to find *anything* uniquely hominid in the capacities needed to make these artefacts.

Finally, the chimpanzee model might usefully be applied even to a recent hominid such as *Homo erectus* with its Acheulean tools. Some of the features put forward as distinctive for these tools (standardisation, symmetry, measurement: see Wynn, 1989) are arguably present in the material culture of living apes. Here the task for investigation is to sort out those abilities that are shared by ape and human from those found only in one or the other.

To some extent, it does not matter which stage in hominisation most closely resembled the chimpanzee or whether or not we can ever specify it. First, stages are only arbitrary segments of what was a continous process. Second, hominisation was a set of messy radiations, not a neat, linear sequence. Third, traits are likely to have evolved as mosaic packages in dynamic compromise, not in racheted progress. Fourth, and most important, pongid material culture has been evolving too, ever since the pongid–hominid split, along its own merry way. What matters more is that chimpanzees are the best non-human species available for modelling the tool-use of our ancestors (McGrew, 1991).

Stone artefacts

Archaeologists seem to be fixated on lithic culture. Received wisdom on the evolutionary origins of technology starts with the recovery of the oldest stones, the positioning or condition of which is thought to be unnatural. This applies whether the earliest signs of lithic technology occur in the Middle Awash (Kalb *et al.*, 1984), Hadar (Harris, 1983), Semliki (Palca, 1986), or wherever. Thus when claims have been made that human-influenced or modified stones have been found at the 'wrong' time or place (Haynes, 1973; Bleed, 1977; Prasad, 1982; Dennell *et al.*, 1988), controversy ensues. The usual pattern is that initially startling claims of greater antiquity (Leakey, 1968) turn out on further scrutiny to be unconvincing (Pickford, 1986).

However, the fixation with stone is entirely understandable since it is mostly what palaeo-archaeologists have to work with. Although items such as bones were originally organic, what is left for study is petrified. This constraint leads to several problems with regard to the behaviour of other animals, especially primates, especially chipanzees. First, the natural activities of many species involve stones, and some of these may produce traces that inadvertently mimic human activity. Second, the manipulation and modification of lithic and non-lithic objects intertwines in primate behaviour, and bias in interpretation may follow if the data are incomplete. Third, some aspects of chimpanzee technology are so similar to that occurring in the palaeo-archaeological record that the two may be indistinguishable. Each of these three points is elaborated below.

Lots of other animals *use* stones. Consider: 'The things that struck us as most remarkable was [*sic*] the unerring judgement in the selection of a pebble of precisely the right size to fit the entrance, and the use of the small pebble in smoothing down and packing the soil over the opening' (Williston, 1892, p. 86). Here a parasitoid wasp uses two stones of differing dimensions in sequential, different tasks (the simplest tool-set?), one as a door and the other as a tamper, in sealing a burrow.

Of more interest here are larger animals that *modify* stones: gastroliths are carried in the guts of animals as diverse as dinosaurs (Wieland, 1907), sea lions (Fleming, 1951) and moas (Smalley, 1979). The stone-swallowers show selectivity in acquiring raw materials and patterned deposition in discarding the polished stones tens of kilometres away from their sources. The most likely candidates for similar activities on African savannas are ostriches; if moas used gizzard-stones of up to 74 millimetres long, then those of ostriches are likely to be even larger. Recently, R. Daigle *et al.* (unpublished data) reported the first primate equivalents: Some rhesus monkeys stuff their cheek-pouches full of stones and keep them there constantly, so that the stones are polished smooth. The function of this bizarre habit is unknown.

Of equal interest are animals that systematically *position* stones, although they do not modify them: several species of African song-birds collect and assemble scores of stones of up to about 30 millimetres diameter as foundations for ground nests (James & Brooke, 1971). Some animals both modify *and* position their raw materials: the mounds of termites on African savannas are mostly composed of uniformly-sized particles of clay. When seasonal bush-fires smoulder for weeks, fragments of the mound may be inadvertently fired to clasts, mistakable for the remnants of early hearths (Gowlett *et al.*, 1981; McGrew, 1989*a*).

Finally, some animals *curate* stones (cf. Binford, 1979), that is, they transport, cache and re-use them. For chimpanzees, this occurs with the stones used as hammers for cracking nuts being carried from one anvil to another in the forest (Boesch & Boesch, 1984*a*). (This is not confined to stones, however, as Gombe's chimpanzees carry vegetation for fishing probes from mound to mound (Goodall, 1968; McGrew, unpublished data).) Chimpanzees may also carry objects over long distances: an adult female at Kasoje carried a piece of pelvic bone for two days, travelling 6 kilometres (Hasegawa *et al.*, 1983). This has interesting implications for conclusions about early hominids transporting lithic raw materials over distances of several kilometres, usually from source to site of use. This inferred activity has been repeatedly cited as evidence of ability to devise and to execute plans (Leakey, 1975; Wynn

& McGrew, 1989), and has been cited as more complex than that of non-humans (Toth & Schick, 1986).

More intriguing is the possibility that such long-distance translocations of stones were not done in one or a few intentional bouts by uniquely curatorial hominids, but instead were the inadvertent, additive by-products of hundreds or thousands of short journeys of only tens of metres by apes. Ongoing computer simulations, using vector analysis and testing against random, 'Brownian' motion patterns suggest that this is the case (McGrew *et al.*, unpublished data). It might take a stone a few years to move a few kilometres by this chimpanzee-like agency, but this is just a moment in evolutionary time.

None of the above examples need have anything to do with hominisation, but all will leave an archaeological record that may cause confusion. This is more than just a hypothetical problem. Chavaillon *et al.* (1979) offered a whole sequence of interpretation starting from Oldowan culture based on 'living-floors' little different from these types of non-human deposits. Leakey (1975) formulated a whole cultural progression at Olduvai on data which shared some of these problems (see below). However, among all living species in nature, only human beings *flake* stone in a functional way (Foley, 1987*b*). Put another way, only human beings make tools with cutting edges.

Among other primates, some stone tool-use is widespread, as in defensive stoning by baboons (Namibia: Hamilton *et al.*, 1975; Sudan: Pettet, 1975; Kenya: Pickford, 1975). In all cases reported the stones were modified in the process ('splintered': Pickford, 1975, p. 549) but apparently only by accident and not design. More complicated and enigmatic is the practice of stone-handling by wild Japanese monkeys (Huffman, 1984; Huffman & Quiatt, 1986). At least three widespread populations spontaneously handled stones in eight ways, and at least four of these patterns (scatter, roll, rub, clack) were likely to modify them. The behavioural patterns are habitual and continue to spread, but no function has yet been divined. Thus one cannot say now whether the monkeys *intended* to alter the stones, but the habit satisfies six of the eight criteria for culture given in Table 4.1. Only diffusion (spread between groups) and naturalness (in unprovisioned groups) remain to be seen. For all wild primates except chimpanzees, use of stone as opposed to perishable organic matter like vegetation is trivial, and their archaeological record will thus be misleadingly sparse (Carbone & Keel, 1985).

For chimpanzees, the balance between lithic and non-lithic technology is similarly biassed. Of the 19 habitual patterns listed in Table 8.2, only four (missile-throw, self-tickle, play-start, nut-hammer) involve stones. Of these, throwing stones as missiles is indistinguishable from similar acts by other primates (see Chapter 7), although this remains to be tested empirically. For self-tickle and play-start, stones seem to be no more preferred than are twigs and other objects.

Only stones used as hammers to crack nuts are undeniably important, and this can be expanded to include stone anvils against which hard-shelled fruits and the skulls of prey are smashed (Boesch & Boesch, 1983, 1989). In neither of these patterns has intentional modification of stone been reported, though hammers may fracture and anvils may be chipped inadvertently. (The same was found for free-ranging chimpanzees on the Bassa Islands, Liberia (Hannah, 1989; Hannah & McGrew, 1987); see Figure 9.2.) However, given the wide range of hammer weights from 1–24 kilograms, it may be that breakage of larger stones produces more efficient smaller ones. Boesch & Boesch (1983) documented the habitual wear patterns, in the form of depressions at the point of impact, produced on hammers and anvils at Tai. Such an indentation in the anvil may increase the efficiency of nut-cracking, as it serves to contain the spherical panda nut on the flat surface of the anvil. Such realignment is clearly intentional, as the nut is precisely and aptly re-positioned. Further, it is often repeated: the mean number of blows needed to crack the hardest nuts is 33 (Boesch & Boesch, 1983).

Future archaeologists excavating an African nut-smashing atelier may be hard-pressed to distinguish human from chimpanzee work-sites. Systematic study is needed at Bossou, where both species crack palm nuts nearby one another (Kortlandt & Holzhaus, 1987; Sugiyama, 1989). Also confusing are sites where only one of the two sympatric species hammers nuts: at Gombe and Kasoje, chimpanzees do not but humans do (D. Collins, personal communication; T. Nishida, personal communication). The local Tanzanian people use a technique to crack palm nuts that seems identical to that of Guinean chimpanzees (see Figure 9.3).

Reconsideration of Table 6.5 shows that only three of 18 subsistants would turn up in the archaeological record for Tasmanians, and only one, used to chop down trees, would show signs of modification. The remaining subsistants were perishable and thus archaeologically invisible (Carbone & Keel, 1985). Table 6.7 shows much the same pattern for chimpanzees: only one of 15 subsistants is of stone, though this excludes anvils that may be modified by repeated use to crack open

Figure 9.2. Close-up of stone hammer and anvil use to crack oil palm nuts, Bassa Islands, Liberia. Note accumulation of nut shells around the anvil. (Courtesy of A. Hannah.)

Figure 9.3. stone hammers and anvil used by local person to crack palm nuts, Gombe, Tanzania (Courtesy of A. Collins.)

hard-shelled fruits. Clearly, relying on lithic technology in the archaeological record is misleading for both ape and human, and might be termed palaeo-myopia (A. Bowes, personal communication).

Why have palaeo-anthropologists ignored other primates?

The 1980s saw a lively, ongoing debate about the daily lives of early hominids, especially in the Plio-Pleistocene of East Africa. There had been such speculation for decades, in an attempt to flesh out a sparse archaeological record, but it was the 'actualistic archaeology' of the late Glynn Isaac and his students that brought a set of controversial issues to the fore. Isaac (1976, 1978*a*, *b*, 1981) posed hypotheses about such key topics as home-bases, food-sharing, butchery, transport, division of labour, ranging, etc., drawing largely on living African hunter-gatherers. His treatment of apes was minimal and simplistic, as illustrated by his contrasting views of chimpanzees and modern humans (see Figure 9.4). Chimpanzees were seen as two-dimensional scroungers. Isaac's framework was used to explain extensive finds of stones-and-bones artefacts from sites in the East African Rift such as Koobi Fora on Lake Turkana in northern Kenya.

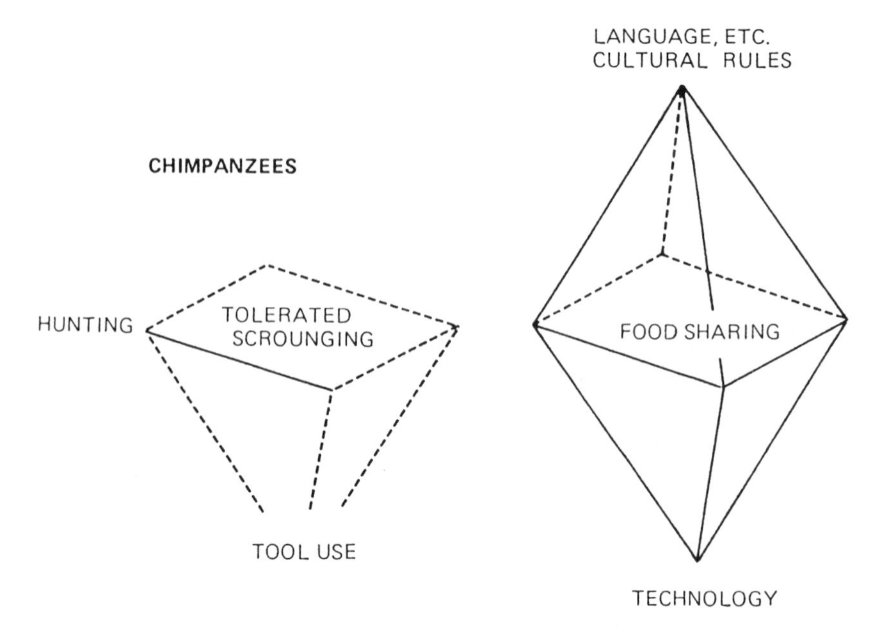

Figure 9.4. Schematic contrast of food-sharing between chimpanzees and modern humans. (After Isaac (1978*a*).)

The debate took the form of reaction and counter-reaction about what could be justifiably inferred from the circumstantial evidence. Alternative explanations for assemblages, especially of associated fauna and artefacts from Olduvai Gorge, focussed on the possibility that natural agents could have been responsible for traces mistakenly attributed to early hominids. The usual alternative agents offered were social carnivores, especially hyaenas, and the most hotly disputed topic has been hunting versus scavenging (Potts, 1984; Binford, 1985; Bunn & Kroll, 1986; Sept, 1986; Shipman, 1986; Toth & Schick, 1986; Blumenschine, 1987; Speth, 1987; Marean, 1989). What is extraordinary in this lively debate is the extent to which advocates virtually ignore the data from other primates in their reconstructions.

The *most* extensive consideration of other primates occurred in Potts's (1988) book-length analyses of material from Olduvai. All non-human primates combined got only five of 311 pages of text, far fewer than carnivores. In the bibliography of almost 400 references, chimpanzees got eight, and meat-eating baboons (Hausfater, 1975) were barely mentioned. Again and again, statements cropped up such as 'hominids and carnivores are clearly the two agents primarily responsible for the bone concentrations at Olduvai' (Potts, 1988, p. 142). Potts could hardly argue that other primates were absent, for more of their fossils were found at the seven sites analysed than there were fossils of hominids. It can be argued that no fossil apes were found at Olduvai, but likewise there were no fossil hominids at five of Potts' seven sites either. Thus, statements such as 'it is now generally believed, though by no means proved, that early *Homo* rather than *Australopithecus* was responsible for the earliest stone tools throughout East Africa' (1988, p. 3) are doubly problematical. Both other hominoids and other (non-*Homo*) hominids were excluded without cause.

So, why have other primates been ignored? One reason may be that other primates truly are irrelevant or inappropriate. Another is that the key primatological knowledge was not accessible, or at least not readily available, to prehistorians. Another is that primatological knowledge is too recent to have worked through to other disciplines. None of these seems to apply.

Consider the following published reports on the habitual activities of wild chimpanzees: They make and use tools (Goodall, 1964). They extract and eat insects (Goodall, 1968). They hunt and eat meat (Teleki, 1973). They share food (McGrew, 1975). They scavenge carcasses from other predators (Morris & Goodall, 1977). They re-use stone tools at work-sites (Sugiyama & Koman, 1979). They process bones with tools (Boesch & Boesch, 1989). All but the last point, plus scores of other findings, were published in mainstream articles, chapters and books *before* the 1980s. What may be telling here is that findings of regular *stone* tool-use by apes did not appear until later (Boesch & Boesch 1981; Sugiyama, 1981). In a nut-shell, ape tool-use has usually been taken seriously by prehistorians only when it involved stone.

Ignorance of other primates applies to many specific, perennial questions in the evolutionary reconstruction of the emergence of humankind. For example, consider the fallacy that *faunivory* equals *carnivory*, that is, that eating animals is the same as eating vertebrates, whether their flesh or other body-parts. Thus ignored is *insectivory*, despite much data on its over-riding importance in the diet of chimpanzees (see Chapters 5 and 7). Almost every published account of the evolution of human diet, whether by ethnologist (Hill, 1982), archaeologist (Speth, 1989) or nutritionist (Eaton & Konner, 1985), omits this basic finding of hominoid natural history, that invertebrates are more dietetically important than vertebrates.

Similarly, in the topical area of interpreting the significance of dental microwear (cf. Grine & Kay, 1988; Walker & Teaford, 1989), tool-use is ignored. It seems questionable to equate dental wear with dietary intake when comparing species that use tools in food processing with species that do not. Chimpanzees using hammers to crack nuts probably eat more hard-shelled food-items than do orang-utans, but their teeth will not reveal it because the hard work is done instead by tools (cf. Teaford & Walker, 1984). So, to try to infer early hominid diet on the basis of dental microwear but without taking account of the ameliorating effects of tool-use (Grine & Kay, 1988; Ryan & Johanson, 1989; Walker & Teaford, 1989) seems even more dubious, if one grants that archaic apes are likely to have had hammer-using skills equivalent to chimpanzees. Why should one hypothesise masticatory adaptations in early hominids for bone-smashing, root-crushing, and nut- or seed-cracking (Peters, 1982) when it is more likely by analogy with living apes that early hominids used tools for these tasks too?

The same point holds if approached from the opposite direction of totally perishable raw materials. Arguably the most pervasive aspect of the material culture of great apes is *nest-building* (the term is ill-chosen, but well-established, cf. Hediger, 1977). Every day of its life after infancy, a great ape makes a sleeping platform or pallet of fresh, usually living vegetation, in which it spends the night (Groves & Sabater Pí, 1985). Often such 'beds' are also made during the day for naps. Each arboreal construction is a skilful inter-weaving of leafy, springy branches, large and small, with a central mattress of twigs and leaflets sometimes detached and added for lining (see Figure 9.5). Terrestrial nests are similarly made, but usually of herbaceous vegetation.

For chimpanzees at least, nests are more than just resting-places (Goodall, 1962, 1968). Many other events, such as birth, copulation, eating, grooming, convalescence and death may take place there. Nest-building is usually not solitary but social, as most chimpanzees sleep in parties, often in the same tree. Although the basic design of nests and patterns of siting are similar species-wide, many differences exist between the nests of different populations: height, openness to the sky, size of nesting party, nests per tree, etc. (Baldwin *et al.*, 1981). Finally, there are marked seasonal differences, especially in location; at the height of the dry season at Assirik, Senegal, a few favoured sites were re-used again and again (Baldwin, 1979; unpublished data).

All of these points have implications for inferring the sleeping habits of early hominids, although the dark half of the tropical 24-hour cycle is

Figure 9.5. Unusually low chimpanzee's nest, Assirik, Senegal.

rarely mentioned in evolutionary scenarios! The reason may be that ape and probably proto-hominid nests are archaeologically invisible, being made entirely of vegetation (Carbone & Keel, 1985). Thus a single surviving circle of stones at Olduvai (Leakey, 1971) is given more prominence in speculation about patterns of habitation than countless lost beds of branches and leaves, which have left no traces (Tuttle, 1988).

An alternative avenue of pursuing sleeping habits of proto-hominids may turn out to be through *coprolites* (Fry, 1985), that is, fossilised faecal deposits, such as Leakey (1971, p. 67) recovered from Upper Bed I at Olduvai (Toth & Schick, 1986). Apes routinely defecate upon arising in the morning and this concentrates their faecal deposits at sleeping sites. Proto-hominids probably did the same.

Some palaeo-anthropologists have made specific use of primatological knowledge, especially in more recent papers (Toth & Schick, 1986; Isaac, 1987; Steele, 1989). However, these are mostly passing, secondary citations. A notable exception is Sept's (1992) work on the archaeological 'visibility' of chimpanzees, based on the distribution of nests at Ishasha, in Zaïre. She makes primary comparisons between primatological data and archaeological data (from Koobi Fora), and concludes that at present the ape and hominid records are indistinguishable. This is exemplary research.

However, the point remains that primatologists have been more keen to link with palaeo-anthropology than vice versa: Kortlandt (1980, 1986), McGrew (1979, 1981*a*, 1991), Suzuki (1975), Teleki (1974, 1975), Wrangham (1987). The solution to the problem may be simple: more inter-disciplinary study in which primatologists and prehistorians work side-by-side doing ethno-archaeology on African primates is needed (Gibson, 1991). Even laboratory collaboration can be fruitful, e.g. in inducing captive apes to make and use stone flakes for cutting (N. Toth & S. Savage-Rumbaugh, unpublished data).

Another cautionary note

Consider the following description: 'My study involved two populations . . . separated by 250 km and morphologically nearly identical . . . The population of the southern Kumawa Mountains erected glued stick towers on a painted black moss base and decorated in stereotyped style with black, brown, and gray snail shells, acorns, sticks, stones, and leaves. The population of the Wandamen Mountains erected woven-stick huts on an unpainted green moss base, decorated with much individual variation, and used fruits, flowers, fungus, and butterfly wings, selecting black plus all of the rainbow colors but making little use of brown, gray, or white' (Diamond, 1988, p. 632).

The two populations (cultures?) referred to are not human, nor even primate, but are bowerbirds! In wooing females, males build elaborate structures termed bowers, which they decorate with a marvellous variety of items. They collect, sometimes by theft, hundreds of objects that they constantly re-arrange in complicated configurations. In some cases they 'paint' the bowers and objects, using a crude brush of crushed leaves. They show marked colour preferences, with blue being a consistent favourite. The bower is meticulously maintained daily and may persist for years. All in all, it is wondrous natural history (Diamond, 1987, 1988).

Bowers are of interest here for several reasons: Demonstrable variation exists not just across species and populations, but across individuals. Young males spend much time watching their adult male neighbours, and it takes 4–7 years for them to develop the local style of adult bowers. Females visit bowers in small groups, and probably acquire proper discrimination by social learning. Apparently arbitrary conventions exist with regard to style: Kumawan birds regularly use acorns as decorations, but Wandamen birds do not, though oaks are the dominant species of tree at both places (Diamond, 1987, p. 199). It would take the broadest treatment of material culture (Lemonnier, 1986), that is, one that takes into account all facets of technical activity, to explain these data, were they presented as human.

Of the variety of raw materials used, the stones will leave an archaeological record; bones, snail shells and pollen may do so, depending on taphonomic conditions; but leaves, petals, fungi, beetle-heads, etc., will be lost. The bower-sites are impressive in size and scale: over 3 square metres of collected material weighing tens of kilograms. This may include hundreds of stones up to 40 millimetres in diameter. Items for bower construction and decoration are brought from as far as 400 metres away. (And, given theft by neighbours, such items may gradually move much farther, bit by bit, so as to be kilometres away from their sources over millennia.) Finally, in addition to creating their own archaeological record, bowerbirds may bias the human one, by their habit of collecting bone, shell and stone (Solomon *et al.*, 1986). They are known to harvest objects from both open and sheltered archaeological sites. It is perhaps just as well that bowerbirds are confined to Australia and New Guinea, and so do not further confuse matters of African prehistory!

The note of caution is that humans or even apes are not the only creatures to leave an archaeological record of a rich array of artefacts.

The complexity of the material culture of the bowerbird was not surpassed until recent evolutionary times, with the appearance of nearly modern humans (cf. Foley, 1987*a*).

Conclusions

Referential modelling may have its drawbacks (Tooby & DeVore, 1987), but the chimpanzee model for hominisation looks better rather than worse, as more knowledge builds up. The gap between human and ape continues to narrow for chimpanzees (e.g. hunting: Boesch & Boesch, 1989), but not for other non-human species. Most importantly, primatological findings continue to pose challenging questions for palaeo-anthropologists to pursue, and this is the most useful test of any model's heuristic value.

10

What chimpanzees are, are not, and might be

> If we, in our travels in space, should encounter a creature that shares 98% of our genetic makeup, think of the money we would spend to study this species. Such creatures exist on earth and we are allowing them to become extinct.
>
> (*Irven DeVore*)

Introduction

Chimpanzees never were, are not now, and probably never will be human beings. The converse is equally true. Yet we and they are sibling species, chromosomally (Yunis & Prakash, 1982) and genetically (Goodman *et al.*, 1990). Some taxonomies place human beings and the African apes in the same subfamily, the Homininae, a classification that would have been unthinkable a generation ago (Groves, 1986). As knowledge accumulates, again and again similarities impress us and force us to abandon cherished clichés of human uniqueness, such as that only human beings intentionally teach their offspring (Boesch, 1991*a*; Boesch & Boesch, 1992). Perhaps the key point is the one that Goodall (1971) has been making for years: *Only when we are clear about the similarities between chimpanzee and human will we be able to recognise the real differences.*

Conceiving of chimpanzees

Accurate interpretation of the capacities of such close relatives as apes is not easy (Jolly, 1991). The two variables are probably inversely correlated: the more like us a species is, the harder (not easier) it is to assess its abilities objectively. These difficulties of comparison take at least four forms:

In *anthropomorphism*, the abilities and motives of other species are over-estimated by interpreting them in human terms. Thus, superficial resemblances are typically endowed with the complex feelings and thoughts that humans have in similar situations. Other species may well have capacities as complex as ours, but this is often impossible to divine with current methods of science. How could we *know* if a chimpanzee was praying? Anthropomorphism often means accepting complicated interpretations when simpler ones will do. Such rich inferences are readily dismissed by invoking the law of *parsimony* (also called Occam's razor, or Lloyd Morgan's canon).

At the other extreme is *speciesism*, in which the capacities of another species are under-estimated on grounds of the presumed superiority of all things human. (Of course, as in all discriminatory 'isms', this process may be unconscious and unintended, as in 'Some of my best friends are animals'.) Speciesism often means denying a complex interpretation of a phenomenon even when no simpler one is advanced. If a chimpanzee shows all the appropriate symptoms of grief, on what grounds can we rule it out?

Both of the above are errors of *anthropocentrism*, which is insisting on viewing the world in human terms. Like ethnocentrism in cross-cultural studies of humans, it may alter even the most mundane perceptions and conceptions of the details of daily life. If eating live insect larvae is repulsive to Westerners, then we may be unable to grant that other cultures or other species may relish such food (Harris, 1985). Similarly if we believe without question that in disciplining children it is sometimes necessary to strike them, we may fail to notice that chimpanzee mothers never do so.

Finally, the least obvious error is *chimpocentrism* (cf. Beck, 1982), in which the perceived similarity of chimpanzees to humans leads to over-estimation of chimpanzees relative to other non-human species. This fallacy is more recent than the others, being a by-product of twentieth-century research, especially the work of Köhler and Goodall. Most primatologists are unaware of chimpocentrism, because of their blinkered state, but among students of other species, entomologists (Hansell, 1987) and mammalogists (Eisenberg, 1973) are notably sensitive. Beck's (1982) comparison of termite-fishing by chimpanzees with mollusc-dropping by gulls is the most detailed treatment of chimpocentrism. He argues that the evidence of underlying mental abilities for the two behavioural patterns is hard to distinguish when compared point by point.

The common thread to these four problems is the question of what can be reasonably inferred about the covert processes, as opposed to the overt acts, of other organisms if we and they cannot communicate directly through verbal disclosure. (And even such communication does not guarantee veracity!) This bedevils issues such as intentionality and consciousness, but these are not unique to other species. The same frustrations apply to pre-verbal infants, post-verbal elders, non-verbal handicapped persons, and allo-verbal members of other cultures. No one baulks at applying the same standards of inference to other cultures of our species, but it is still easy to move the goalposts when another species is involved (Jolly, 1991).

Nowhere is this more marked than in the attribution of culture. When differences emerge between human groups it is *assumed* that these are cultural. No one's first explanation of (for example) a person's avoidance of touching food with the left hand would be that it is natural. Yet what would we make of exactly the same behavioural pattern if shown by apes? Consider a gradation in some aspect of material culture such as the design of throwing sticks across a continent (Oswalt, 1973) or of ear ornaments within a region (Hodder, 1977) that emerges ethnographically. Whether or not the gradation results from diffusion may be debated, but no one questions that it is cultural. What about the same sort of data, but for apes?

For example, suppose we found a population of previously unknown hominids who avoided contact with surface water. Not only did they not swim in lakes, paddle in ponds, or ford rivers, but they never waded across the shallowest and narrowest streams and even detoured around puddles on paths. If we perceived of these creatures as human-like, we might try to explain the act in terms of custom, tradition, ritual, or even symbolic taboo. If we perceived of them as ape-like, we might think of the behaviour as instinctive, hard-wired, species-typical, adaptive, etc. In fact, the chimpanzees at Gombe show just this reticence (McGrew, 1977) while chimpanzees at Kasoje do not (Nishida, 1980*a*).

Evolutionarily relevant gaps

Chapter 6 contrasted chimpanzees and living foraging peoples in terms of diets and food acquisition and processing. Here the comparisons are broadened and reprised, and then extended to other aspects of daily life. The emphasis is on what evolutionary steps would fill the gaps between human and ape (McGrew, 1991).

Hunting

Both species decide to hunt prey, and then seek, stalk, pursue, capture and dispatch it. The first stage is hardest to discern, given the apparently non-verbal nature of chimpanzee communication, while humans may (or may not) practise elaborate pre-hunting routines, including rites. For chimpanzees, Boesch & Boesch (1989) have distinguished opportunistic from intentional hunting on empirically testable grounds.

Both species search for prey in response to visual, vocal or olfactory spoor. Both use searching *tactics*, such as visual scanning from an elevated site, but it is not clear whether apes use searching *strategies*. For us to know whether or not chimpanzees systematically seek out places likely to be used by bushbucks to cache their fawns would require more data on the ecology of the prey species. Both humans and apes stalk prey, but it is unclear whether apes use strategies of concealment to approach or to ambush prey.

Humans use various weapons to capture and to dispatch vertebrate prey while chimpanzees do not. Few of the weapons, implements, facilities, etc., categorised by Oswalt (1973, 1976; see Chapter 6) are used by apes, except in isolated cases (Plooij, 1978*b*). This may be the single most important reason why chimpanzees do not take larger prey, but it is puzzling that chimpanzees do not club porcupines, prise out bushbabies, or flail at monitor lizards. (J. Hart's (personal communication) preliminary evidence of sticks to open tortoises remains tantalisingly ambiguous.)

Finally, both humans and chimpanzees hunt solitarily or socially, but only recently has an operational framework been devised that allows comparison across species. Boesch & Boesch (1989) defined four levels of increasing complexity in cooperative hunting: *similarity*, *synchrony*, *coordination*, *collaboration*. With reference to possible chimpocentrism, it may be that in all stages of hunting, social carnivores such as African hunting dogs are just as much like humans as are chimpanzees. This remains to be tested. However, no canids, felids or hyaenids use tools in any form at any stage of their hunting!

Gathering

Humans and chimpanzees may directly compete for most plant parts, especially fruit, but the two main sources of carbohydrates common in human diet are notably ignored by chimpanzees: underground storage organs (tubers) and seeds of grasses (cereals). Obtaining big-enough roots is both time-consuming and energy-consuming unless heavy-duty, special-purpose digging tools are used (Vincent, 1984). Such tools are easily made with flaked stone tools and fire (Sussman, 1986), but may be too costly to be made with teeth alone. Also, prolonged activity at a terrestrial site in the open, where most suitable tubers are found, may make the diggers vulnerable to predators. This argument applies even more strongly to cereals, which tend to occur in broad, often single-species swathes in biomes where trees are few and small. Further, grass seeds though nutritious and accessible, are small and picky to handle without agricultural techniques.

Thus tubers and cereals make special ecological and technological demands that human beings but not apes have solved. Most chimpanzees do not occupy, and so have not been studied in, the dry, open, highly seasonal environments where these two plant parts flourish. Savannas are ecosystems in which terrestrial-travelling and often solitary creatures like chimpanzees would be vulnerable to social carnivores.

As is often the case (Beck, 1974) the baboon's solution is instructive: foraging *en masse* allows safer access to the open spaces that offer small underground items such as corms or rhizomes. These can be uprooted or dug up quickly by hand then, along with grass seeds, collected in cheek-pouches for later consumption at a safer place (Rhine & Westlund, 1978). Thus baboons use natural containers, as do ruminants. (It is worth noting that the only grazing species of primate, the gelada of the alpine meadows of Ethiopia, faces no natural terrestrial predators (Wrangham, 1980a).) Custom-made excavators and containers and anti-predator weapons comprise an adaptive suite for gathering that is uniquely human, but apes make or will use all the components in isolation.

Food processing

As shown earlier, humans and apes can be very similar in some extractive techniques, such as using a stone as a hammer to crack nuts (see Figure 10.1). However, there the resemblance ends, in that only humans use tools to *transform* their food, while apes eat it as it comes. Ape food processing extends only to dismantling it. Chief among the human transformations is heating (Stahl, 1984), which entails the use and usually the control of fire. Other human processing techniques

Figure 10.1. cracking oil palm nuts with stone hammer and wooden anvil, Bassa Islands, Liberia. (Courtesy of A. Hannah.)

involve substituting tools for body-parts in pounding, grinding, scraping, soaking, etc. All of these can be done with simple naturefacts, and the actions used fall within the behavioural capacities of apes, but most also involve containers. Chimpanzees in captivity readily and spon-

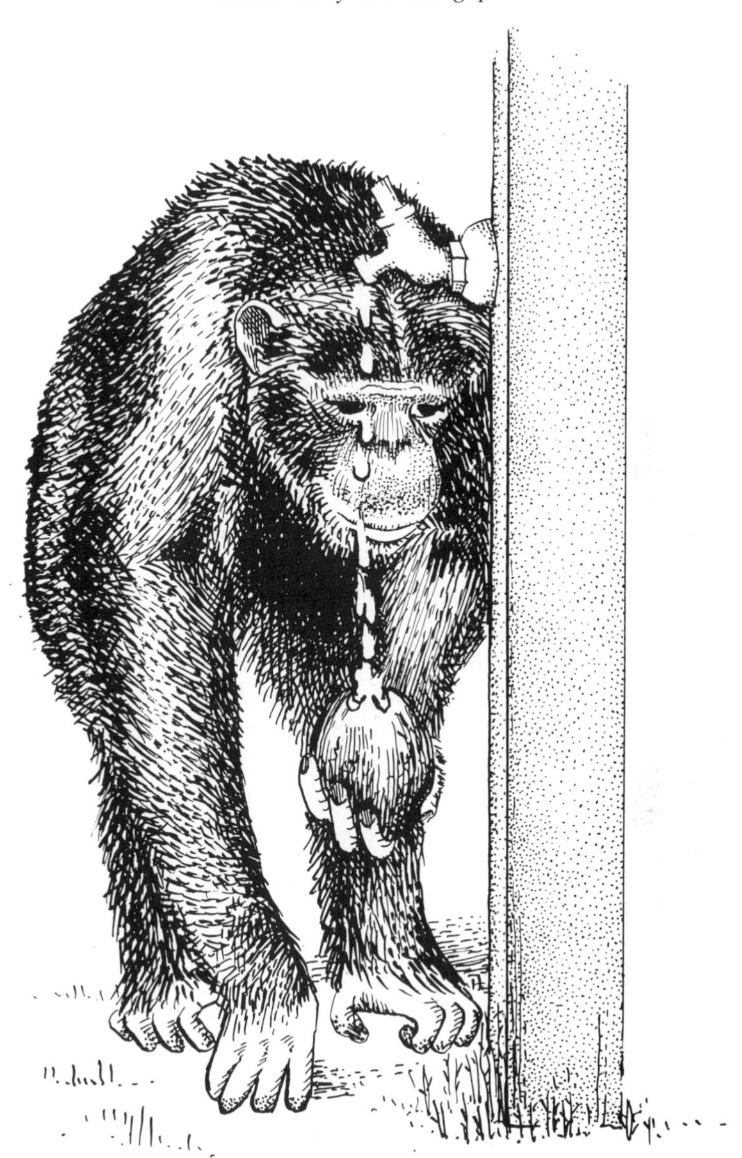

Figure 10.2. Use of empty coconut shell as container to collect water, Delta Primate Center, USA. (Courtesy of B. Merrick.)

taneously make use of containers (McGrew & C. Tutin, unpublished data; see Figure 10.2) and understand their principles (Woodruff & Premack, 1978), but do not use them in nature. More demanding is butchery, which requires a sharp edge, at least for larger prey that cannot be torn to bits. No chimpanzee tool used for any purpose makes use of a sharp edge.

Food is only minimally transported and stored by chimpanzees. Without containers or cheek pouches or sacculated guts, apes move food only by carrying it in hand, foot or mouth, by draping it over the shoulders or by tucking it in the groin 'pocket'. All of these alternatives allow transfer of only small amounts or over short distances. Moving bipedally increases the carrying capacity of the upper limbs, but at the expense of loss of efficiency at high speeds, as when fleeing from predators (Rodman & McHenry, 1980).

Storage is even more dependent on containers, and seems to be non-existent in free-ranging chimpanzees. Non-human primates may hoard, but only in captivity where food is less prone to usurpation by pests (Marriott & Salzen, 1979). Prey animals taken in hunting by chimpanzees are not big enough to justify stashing them for repeated meals, in the way that other predators do (Cavallo & Blumenschine, 1989). In any event, most chimpanzees' foods are highly perishable, so that attempted storage would be useless or even dangerous (Janzen, 1977). The exception is nuts, which are both high in quality and 'pre-packaged' for transport and storage. If there is anywhere to expect wild chimpanzees to use containers, it is in the forests of far western Africa where nut-cracking is common.

Communication

Apart from early attempts by Marler and his colleagues (Marler & Hobbett, 1975), natural communication by chimpanzees has been curiously neglected (Snowdon, 1990), although there are recent signs of a revival of interest (Boehm, 1989; Hauser & Wrangham, 1987). Instead most research has concentrated on the willingness and ability of apes to make use of human communicatory systems, either habitual or devised (Gardner & Gardner, 1969; Premack, 1971; Rumbaugh, 1977; Savage-Rumbaugh, 1986). Many of the protocols in 'pongo-linguistics' qualify as aspects of *material* culture, as they make use of objects such as tokens or keyboards. Further, even when the communication is purely gestural, as in sign language, its referents are usually concrete items such as food or toys (nouns) or the actions of use of such items (verbs).

Some studies have specifically used tools as items, such as in the context of problem-solving (Savage-Rumbaugh *et al.*, 1978). However, all of the pongo-linguistic research seems tightly bounded, in that the ape subjects have little or no chance to work creatively, such as to make tokens or to combine keyboard elements. Thus it is hard to draw comparisons across tool-*making* and communication, except for the occasional spontaneous inventions of signs by sign-language-using apes, about which the evidence remains anecdotal.

The relative lack of study of natural communication by apes means that even basic processes remain unknown (Snowdon, 1990). For example, we do not know whether chimpanzees can identify one another's drumming signatures, though drumming is a species-typical component of displays. It remains unclear whether any chimpanzee communication is truly cooperative, or is merely coordinated. The distinction here is between selfish acts performed in parallel by two or more individuals at once (*coordination*) and collective action that produces a greater *pro rata* pay-off for all participants (*cooperation*). For example, chimpanzee males patrol and display together in the maintenance of territorial relations with their neighbours. Such mass action may just be individuals acting in concert, so that the effect of the display is merely the sum of the number and identities of the participants. Or, cooperating groups of displayers may divide the labour, such as 'You drum while I hoot' or 'Let's take turns calling so as to keep it going longer'. Such data remain to be collected, but scrutiny of cooperation in other aspects of daily life such as hunting is promising (Boesch & Boesch, 1989).

Several things need doing: Field-work on chimpanzee communication should be undertaken by linguists in collaboration with ethologists (Gibson, 1991). Studies of reciprocity should widen their scope to include the *signals* used to initiate, maintain and terminate exchanges, as in food sharing. Menzel's earlier studies (1974) at the Delta Primate Center suggested the existence of such phenomena but did not go on to specify or to analyse them. At this point we know more about the natural communication of vervet monkeys than of any ape (Cheney & Seyfarth, 1990). We cannot now say how much chimpanzees can *negotiate*, such as 'How many minutes of social grooming must I invest before you will let me use your hammer-stone?' or 'How many pieces of meat must I give to you before you will give me priority in mating next time you are in oestrus?' We need to know how good chimpanzees are as accountants.

Unanswered questions

The preceding section should scotch any suggestions that our knowledge of chimpanzees is complete. There may be a tendency to think that after

Figure 10.3. Laterality of function in fishing for termites, Gombe, Tanzania: right hand inserts tool while left hand gives postural support.

80 years of laboratory study and 30 years of field-study chimpanzees are 'done'. This is not so, and the aim of this section is to pose or at least to remind readers of questions that remain to be answered, or in some cases even to be addressed. The focus here remains on aspects of material culture and all of the examples raised below are of direct relevance to hominisation.

For example, it remains unknown to what extent chimpanzees show *laterality of function* such as hand preference or handedness (MacNeilage *et al.*, 1987). There are few satisfactory data on this from field-studies of any apes (cf. Byrne & Byrne, 1991). For tool-use tasks practised by wild chimpanzees there are only two reports that go beyond limited categorisation and data (Nishida, 1973; Nishida & Hiraiwa, 1982). The exceptions are Boesch's (1991*a*) study of hand use in nut-cracking at Tai, and McGrew & Marchant's (1992) analysis of hand use in termite fishing at Gombe (see Figure 10.3). Both found marked individual hand preferences, but no population-level handedness; this concurs with results from tool-use by captive chimpanzees (Steiner, 1990; Marchant, 1992).

We cannot yet say whether there is a correlation between extent or type of laterality and such variables as age, sex, grip, posture, tool characteristics, task requirements, etc. One might predict a greater likelihood of laterality in a two-handed, complementary behavioural pattern such as ant-dipping than in a one-handed pattern such as nut-cracking, but this remains to be seen (see Figure 10.4). Laterality in other sensorimotor functions, such as eyedness, earedness or footedness, has been studied only minimally, in either field or laboratory (Falk, 1987; Marchant & McGrew, 1991). When more knowledge is available from living hominids, inferences about the evolutionary emergence of hominid laterality of function will be strengthened (Toth, 1985*a*).

In the related area of motor skills, the distinction between aimed and unaimed throwing of missiles emerged early (Kortlandt & Kooij, 1963; Goodall, 1964). However, apart from gross treatment (Marchant, 1992), it has not been followed up systematically or even operationally defined (McGrew, 1992*b*) (see Figure 10.5). Underarm versus overarm, bipedal versus tripedal, one-handed versus two-handed, etc., are all obvious independent variables for study, while distance, accuracy, trajectory, etc., are obvious dependent variables. Several authors (Isaac, 1987) have asserted the crucial importance of throwing in human evolution, even to the point of reification (Calvin, 1982), but this is little more than speculation. If behaviourists can teach monkeys in zoos to play reaction time games (Markowitz & Spinelli, 1986), surely apes can be induced to throw, given suitable rewards.

Many unanswered questions about chimpanzee material culture fall in the area of ethno-archaeology. Consider meat-eating. Despite detailed observations on some aspects of the sequence (Teleki, 1973) it remains unknown what remains after consumption. No one has collected and described these remnants, nor their distribution in time and space. In the fossil record, palaeo-anthropologists would not recognise a bone chewed by an ancestral ape because we have not bothered to look at bones from kills by living apes (McGrew, 1992*b*). We do not know whether or how chimpanzees could use cutting edges or piercing points in subsistence tasks such as butchery. In captivity, it would be easy enough to induce or to allow chimpanzees to make picks of bone (Boesch & Boesch, 1989) or flakes of stone (Wright, 1972; N. Toth & S. Savage-Rumbaugh, unpublished data), and then present them with carcasses in differing states.

Several key ecological questions will be answered only if and when

Figure 10.4. Laterality of function in dipping for driver ants, Gombe, Tanzania: (*a*) left-hand inserts tool while right hand gives support; (*b*) left foot holds tool, left hand does pull-through, and right hand gives full suspensory support. (Courtesy of D. Bygott.)

chimpanzees living on savannas are habituated to observation at close range and for long periods. Potentially this could be done at least in Senegal, Tanzania or Mali, though intervention is probably needed. Advances in micro-electronics and satellite communications now enable easier and more efficient telemeterised location and monitoring of wide-ranging subjects. We do not know the extent to which such chimpanzees scavenge carcasses or how they do so, such as their tactics for competing with other carnivores. Experimental intervention could be done easily, once subjects can be followed to their overnight sleeping sites. Stealthy deposition nearby of a carcass in the middle of the night could be followed by recording the responses of the apes the next morning (cf. Kortlandt, 1967).

Figure 10.4 *cont.*

On the psychological front, basic questions remain unanswered about the social transmission of information. Do chimpanzees *teach*, that is, does one chimpanzee act in such a way as to cause or to enhance the acquisition of knowledge or skill in another? It is hard to distinguish such tuition from inadvertently acting as a demonstrator for an onlooker while going about the activities of daily life (cf. Boesch & Boesch, 1992) (see Figure 10.6). In its simplest form, teaching may simply be scheduling performances so that the pupil is present and attentive (Boesch, 1991*a*). Another form of teaching is selective interference in the activities of another, which may gradually shape the pupil's behaviour. Such acts have been seen occasionally in wild chimpanzees (Nishida, 1983), but do not occur regularly, even in detailed studies of mother–infant interaction (Plooij, 1984).

More complex teaching could be pursued in experimental settings in

Figure 10.5. Adolescent males throws rock underarm at observer overhead, Delta Primate Center, USA.

Figure 10.6. Infant watches closely while her mother fishes for termites, Gombe, Tanzania. (Courtesy of C. Tutin.)

which the tutor's reward is made dependent on the tutee's acquisition of knowledge or skills. Studies have been done on cooperative tool-use (Beck, 1973; Savage-Rumbaugh *et al.*, 1978), but the processes whereby the cooperation was learned, and whether or not this involved teaching or trial-and-error discovery, remain unknown. Anecdotes abound of teaching-like acts by apes, but most of these involve humans and apes in testing sessions (Hayes & Hayes, 1952), and the challenge is to arrange a context in which teaching will occur spontaneously and unconfounded, ape to ape (cf. Fouts, 1989). In any event, many arguments about culture are ultimately predicated on the issue of proven intentionality (Ingold, 1986*c*), and if teaching as defined above occurs in apes, a major distinction between humans and non-humans would disappear. (See also Carrither's (1990) distinction between chimpanzee *training* and human *pedagogy*, which unfortunately lacks an operational distinction.)

One topic that combines several of the above aspects is the use of *fire*. Chapter 6 gave preliminary data from both captivity and nature that suggest that chimpanzees may use fire if given the chance. To what extent they may control or even make fire remains unknown, but both points are pertinent to the debate about the advent of domesticated fire in human evolution (Gowlett *et al.*, 1981; Clark & Harris, 1985; Goudsblom, 1986; James, 1989). Studies of free-ranging, savanna-dwelling chimpanzees should concentrate on their activities at times when bush-fires sweep through their ranges (McGrew, 1989*a*). This usually occurs at the beginning of the dry season, and again supplemental field experimentation could be informative, such as the nocturnal presentation of raw versus roasted plant and animal foods near sleeping sites.

In captivity, chimpanzees could be given the opportunity to use, maintain, re-kindle, and even ignite fire in controlled (and safe!) settings. (Orang-utans being rehabilitated readily acquire such skills, if given access to appropriate objects and materials (Russon & Galdikas, 1992).) Its spontaneous use for heat, light, cooking and defence could be studied. Would chimpanzees sleep near fires on cold nights? Use torches to illuminate dark places? Roast tough pods to dehisce seeds? Incorporate brands as weapons? Stoke a fire in response to the broadcast roars of lions in the dark? Fan coals to revive them? Use one fire to light another? Such questions may verge on the fantastic, but answering them is feasible, if someone can be bothered to do the research.

Conclusions

Chimpanzees do not have human culture, material or otherwise. Similarly, even the simplest aspects of human culture are not those of apes, or other primates, mammals or vertebrates. Yet much of what chimpanzees do is so close to human that the two are indistinguishable. Some artefacts would be unattributable to species if they lost their museum labels. This similarity is of more than academic interest, for it is the best available source of knowledge about our behavioural evolutionary past. If we wish to reconstruct the prehistoric origins of human technology, then we need to use the available acts of the creatures with whom we last shared a common ancestor. Our hominid predecessors are irretrievably gone, but our hominoid cousins (just) survive. What a pity it would be to extinguish them before they could tell us all that they know.

Appendix. Scientific names

Common and scientific names of animals and plants mentioned in the text, when known. When there is no common name in English, the scientific name is used in the text. The most specific name is given, to the most precise taxon (see also Table 3.1).

Common name	*Scientific name*
Aardvark	*Orycteropus afer*
Acorn woodpecker	*Melanerpes formicivorus*
African hunting dog	*Lycaon pictus*
Ants	Formicidae, Hymenoptera
Baboons	*Papio* spp.
Bananas	*Musa* spp.
Baobab	*Adansonia digitata*
Bees and wasps	Hymenoptera
Blue duiker	*Cephalophus monticola*
Blue monkey	*Cercopithecus mitis*
Bonobo (pygmy chimpanzee)	*Pan paniscus*
Bowerbirds	Ptilonorhynchidae
Bugs	Hemiptera
Bushbaby	*Galago senegalensis*
Bushbuck	*Tragelaphus scripta*
Bush pig	*Potamochoerus porcus*
Butterflies and moths	Lepidoptera
Capuchin monkey	*Cebus apella*
Central-western chimpanzee	*Pan t. troglodytes*
Chimpanzee	*Pan troglodytes*
Crows	*Corvus* spp.
Driver ant	*Dorylus (Anomma)* spp.
Ducks	Anatidae
Eastern chimpanzee	*Pan t. schweinfurthii*
Eastern lowland gorilla	*Gorilla g. graueri*
Far western chimpanzee	*Pan t. verus*

Common name	*Scientific name*
Figs	*Ficus* spp.
Fig wasps	*Blastophaga* spp.
Flies	Diptera
Francolins	*Francolinus* spp.
Gelada baboon	*Theropithecus gelada*
Gibbons	*Hylobates* spp.
Gorilla	*Gorilla gorilla*
Great Apes	Pongidae
Guinea baboon	*Papio papio*
Highland (mountain) gorilla	*Gorilla g. beringei*
Honey badger	*Mellivora capensis*
Honey-bee	*Apis mellifera*
Human being (modern)	*Homo sapiens sapiens*
Japanese macaque	*Macaca fuscata*
Kangaroos	Macropodidae
Leopard	*Panthera pardus*
Lion	*Panthera leo*
Moa	Dinornithiformes
Monitor lizards	*Varanus* spp.
Mongongo nut	*Ricinodendron rautanenii*
Mound-building termites	Macrotermitinae
Oil palm	*Elaeis guineensis*
Olive baboon	*Papio anubis*
Orang-utan	*Pongo pygmaeus*
Ostrich	*Struthio camelus*
Parasitoid wasps	*Ammophila* spp.
Porcupines	Hystrichidae
Red colobus monkey	*Colobus badius*
Red-tailed monkey	*Cercopithecus ascanius*
Rhesus macaque	*Macaca mulatta*
Sea lions	*Zalophus* spp.
Siamang	*Symphalangus syndactylus*
Striped tree squirrel	*Funisciurus* sp.
Termites	Isoptera
Tits	*Parus* spp.
Tortoises	Testudinae
Tree pangolin	*Manus tricuspis*
Vervet monkey	*Cercopithecus aethiops*
Warthog	*Phacochoerus aethiopicus*
Weaver ant	*Oecophylla longinoda*
Weaver bird	*Ploceus* spp.
Western lowland gorilla	*Gorilla g. gorilla*
White-collared mangabey	*Cercocebus torquatus*

References

Adang, O. M. J. 1986. Exploring the social environment: a developmental study of teasing in chimpanzees. *Ethology*, **73**, 136–60.
Albrecht, H. 1976. Chimpanzees in Uganda. *Oryx*, **13**, 357–61.
Albrecht, H. & Dunnett, S. C. 1971. *Chimpanzees in Western Africa*. Munich: Piper.
Altmann, J. 1974. Observational study of behavior: sampling methods. *Behaviour*, **49**, 227–67.
Altmann, S. A. & Altmann, J. 1977. On the analysis of rates of behaviour. *Animal Behaviour*, **25**, 364–72.
Anderson, A. R. & Moore, O. K. 1962. Toward a formal analysis of cultural objects. *Synthese*, **14**, 144–70.
Anderson, J. R. 1990. Use of objects as hammers to open nuts by capuchin monkeys (*Cebus apella*). *Folia Primatologica*, **54**, 138–45.
Anderson, J. R., Williamson, E. A. & Carter, J. 1983. Chimpanzees of Sapo Forest, Liberia: density, nests, tools and meat-eating. *Primates*, **24**, 594–601.
Anonymous, 1971. Siamang also a tool user. *Yerkes Newsletter*, **8**, 12.
Anzenberger, G. 1977. Ethological study of African carpenter bees of the genus *Xylocopa* (Hymenoptera, Anthophoridae). *Zeitschrift für Tierpsychologie*, **44**, 337–74.
Azuma, S. & Toyoshima, A. 1961–2. Progress report of the survey of chimpanzees in their natural habitat, Kabogo Point area, Tanganyika. *Primates*, **3**, 61–70.
Badrian, A. & Badrian, N. 1977. Pygmy chimpanzees. *Oryx*, **13**, 463–8.
Badrian, N. & Malenky, R. K. 1984. Feeding ecology of *Pan paniscus* in the Lomako Forest, Zaïre. In: *The Pygmy Chimpanzee*, ed. R. L. Susman, pp. 275–99. New York: Plenum Press.
Bahuchet, S. 1978. Les constraintes écologiques en forêt tropicale humide: L'exemple des Pygmées Aka de la Lobaye (Centrafique). *Journal d'Agriculture Tropicale et de Botanique Apliquée*, **25**, 1–29.
Bailey, R. C., Head, G., Jenike, M., Owen, B., Rechtman, R. & Zechenter, E. 1989. Hunting and gathering in tropical rain forest: is it possible? *American Anthropologist*, **91**, 59–82.
Bailey, R. C. & Peacock, N. R. 1988. Efe Pygmies of northeast Zaïre: subsistence strategies in the Ituri forest. In: *Coping with Uncertainty in Food Supply*, ed. I. de Garine and G. A. Harrison, pp. 88–117. Oxford: Clarendon Press.

Baldwin, L. A. & Teleki, G. 1976. Patterns of gibbon behavior of Hall's Island, Bermuda. A preliminary ethogram for *Hylobates lar*. In: *Gibbon and Siamang*, vol. 4, ed. D. M. Rumbaugh, pp. 21–105. Basel: S. Karger.
Baldwin, P. J. 1979. The natural history of the chimpanzee (*Pan troglodytes verus*) at Mt. Assirik, Senegal. PhD thesis, University of Stirling.
Baldwin, P. J., McGrew, W. C. & Tutin, C. E. G. 1982. Wide-ranging chimpanzees at Mt. Assirik, Senegal. *International Journal of Primatology*, **3**, 367–85.
Baldwin, P. J., Sabater Pí, J., McGrew, W. C. & Tutin, C. E. G. 1981. Comparisons of nests made by different populations of chimpanzees (*Pan troglodytes*). *Primates*, **22**, 474–86.
Beatty, H. 1951. A note on the behavior of the chimpanzee. *Journal of Mammalogy*, **32**, 118.
Beck, B. B. 1967. A study of problem solving by gibbons. *Behaviour*, **28**, 95–109.
Beck, B. B. 1973. Cooperative tool use by captive hamadryas baboons. *Science*, **182**, 594–7.
Beck, B. B. 1974. Baboons, chimpanzees, and tools. *Journal of Human Evolution*, **3**, 509–16.
Beck, B. B. 1977. Kohler's chimpanzees: how did they really perform? *Zoologische Garten*, **5**, 352–60.
Beck, B. B. 1980. *Animal Tool Behavior*. New York: Garland STPM Press.
Beck, B. B. 1982. Chimpocentrism: bias in cognitive ethology. *Journal of Human Evolution*, **11**, 3–17.
Benedict, R. 1935. *Patterns of Culture*. London: Routledge and Kegan Paul.
Benson, C. W. 1968. An alleged record of the chimpanzee *Pan satyrus* in Malawi. *Society of Malawi Journal*, **21**, 7–12.
Bermejo, M., Illera, G. & Sabater Pí, J. 1989. New observations on the tool-behavior of the chimpanzees from Mt. Assirik (Senegal, West Africa). *Primates*, **30**, 65–73.
Binford, L. R. 1979. Organization and formation processes: looking at curated technologies. *Journal of Anthropological Research*, **35**, 255–73.
Binford, L. R. 1985. Human ancestors: changing views of their behavior. *Journal of Anthropological Archaeology*, **4**, 292–327.
Birch, H. G. 1945. The relation of previous experience to insightful problem-solving. *Journal of Comparative Psychology*, **38**, 367–83.
Bird-David, N. 1990. The giving environment: another perspective on the economic system of gatherer-hunters. *Current Anthropology*, **31**, 189–96.
Bleed, P. 1977. Early flakes from Sozudai, Japan: are they man-made? *Science*, **197**, 1357–9.
Bloch, M. 1991. Language, anthropology and cognitive science. *Man*, **26**, 183–98.
Blumenschine, R. J. 1987. Characteristics of an early hominid scavenging niche. *Current Anthropology*, **28**, 383–407.
Blumenschine, R. J. & Selvaggio, M. M. 1988. Percussion marks on bone surfaces as a new diagnostic of hominid behaviour. *Nature*, **333**, 763–5.
Boehm, C. 1989. Methods for isolating chimpanzee vocal communication. In: *Understanding Chimpanzees*, ed. P. G. Heltne and L. A. Marquardt, pp. 38–59. Cambridge, Mass.: Harvard University Press.
Boesch, C. 1978. Nouvelles observations sur les chimpanzés de la forêt de Tai (Côte-d'Ivoire). *La Terre et la Vie*, **32**, 195–201.

Boesch, C. 1991*a*. Teaching among wild chimpanzees. *Animal Behaviour*, **41**, 530–2.
Boesch, C. 1991*b*. Handedness in wild chimpanzees. *International Journal of Primatology*, **12**, 541–80.
Boesch, C. & Boesch, H. 1981. Sex differences in the use of natural hammers by wild chimpanzees: a preliminary report. *Journal of Human Evolution*, **10**, 585–93.
Boesch, C. & Boesch, H. 1983. Optimisation of nut-cracking with natural hammers by wild chimpanzees. *Behaviour*, **83**, 265–86.
Boesch, C. & Boesch, H. 1984*a*. Mental map in wild chimpanzees: an analysis of hammer transports for nut cracking. *Primates*, **25**, 160–70.
Boesch, C. & Boesch, H. 1984*b*. Possible causes of sex differences in the use of natural hammers by wild chimpanzees. *Journal of Human Evolution*, **13**, 415–40.
Boesch, C. & Boesch, H. 1989. Hunting behavior of wild chimpanzees in the Tai National Park. *American Journal of Physical Anthropology*, **78**, 547–73.
Boesch, C. & Boesch, H. 1990. Tool use and tool making in wild chimpanzees. *Folia Primatologica*, **54**, 86–99.
Boesch, C. & Boesch, H. 1992. Transmission aspects of tool use in wild chimpanzees. In: *Tools, Language and Intelligence: Evolutionary Implications*, ed. T. Ingold and K. R. Gibson. Oxford: Oxford University Press.
de Bournonville, C. 1967. Contribution à l'étude du chimpanzé en République de Guinée. *Bulletin de l'Institut Fondamental d'Afrique Noire*, **29A**, 1188–269.
Boysen, S. T. & Berntson, G. G. 1989. Numerical competence in a chimpanzee (*Pan troglodytes*). *Journal of Comparative Psychology*, **103**, 23–31.
Brain, C. K., Churcher, C. S., Clark, J. D., Grine, F. E., Shipman, P., Susman, R. L., Turner, A. & Watson, V. 1988. New evidence of early hominids, their culture and environment from the Swartkrans cave, South Africa. *South African Journal of Science*, **84**, 828–35.
Brewer, S. M. 1978. *The Chimps of Mt. Asserik*. New York: Knopf.
Brewer, S. M. 1982. Essai de réhabilitation au Parc National du Niokolo-Koba de chimpanzés auparavant en captivité. *Mémoires de l'Institut Fondamental d'Afrique Noire*, **92**, 341–62.
Brewer, S. M. & McGrew, W. C. 1990. Chimpanzee use of a tool-set to get honey. *Folia Primatologica*, **54**, 100–4.
Brink, A. S. 1957. The spontaneous fire-controlling reactions of two chimpanzee smoking addicts. *South African Journal of Science*, **53**, 241–7.
Brown, J. K. 1970. A note on the division of labour by sex. *American Anthropologist*, **72**, 1073–8.
Bunn, H. T., Bartram, L. E. & Kroll, E. M. 1988. Variability in bone assemblage formation from Hadza hunting, scavenging, and carcass processing. *Journal of Anthropological Archaeology*, **7**, 412–57.
Bunn, H. T. & Kroll, E. M. 1986. Systematic butchery by Plio/Pleistocene hominids at Oldovai Gorge, Tanzania. *Current Anthropology*, **27**, 431–52.
Burton, R. [1966]. *A Mission to Gelele King of Dahomey*. London: Routledge and Kegan Paul.

Busse, C. B. 1977. Chimpanzee predation as a possible factor in the evolution of red colobus monkey social organization. *Evolution*, **31**, 907–11.

Butynski, T. M. 1982. Vertebrate predation by primates: a review of hunting patterns and prey. *Journal of Human Evolution*, **11**, 421–30.

Butynski, T. M. 1986. Primates and their conservation in the Impenetrable Forest, Uganda. *Primate Conservation*, **6**, 68–72.

Byrne, R. W. & Byrne, J. M. 1991. Hand preferences in the skilled gathering tasks of mountain gorillas (*Gorilla g. berengei*). *Cortex*, **27**, 521–46.

Byrne, R. W. & Whiten, A. (eds.) 1988. *Machiavellian Intelligence: Social Expertise and the Evolution of Intellect in Monkeys, Apes and Humans*. Oxford: Clarendon Press.

Caccone, A. & Powell, J. R. 1989. DNA divergence among hominoids. *Evolution*, **43**, 925–42.

Cafagna, A. C. 1960. A formal analysis of definitions of 'culture'. In: *Essays in the Science of Culture in Honor of Leslie A. White*, ed. G. E. Dole and R. L. Carneiro. New York: Thomas Y. Crowell.

Calvin, W. H. 1982. Did throwing stones shape hominid brain evolution? *Ethology and Sociobiology*, **3**, 115–24.

Carbone, V. A. & Keel, B. C. 1985. Preservation of plant and animal remains. In: *The Analysis of Prehistoric Diets*, ed. R. I. Gilbert and J. H. Mielke, pp. 1–19. Orlando: Academic Press.

Caro, T. M., Roper, R., Young, M. & Dank, G. R. 1979. Inter-observer reliability. *Behaviour*, **69**, 303–15.

Carpenter, C. R. 1937. An observational study of two captive mountain gorillas (*Gorilla beringei*). *Human Biology*, **9**, 175–96.

Carrithers, M. 1990. Why humans have cultures. *Man*, **25**, 189–206.

Carroll, C. R. 1979. A comparative study of two ant faunas: the stem-nesting ant communities of Liberia, West Africa and Costa Rica, Central America. *American Naturalist*, **113**, 551–61.

Carroll, R. W. 1986. Status of the lowland gorilla and other wildlife in the Dzanga-Sangha region of southwestern Central African Republic. *Primate Conservation*, **7**, 38–41.

Carroll, R. W. 1988. Relative density, range extension, and conservation potential of the lowland gorilla (*Gorilla gorilla gorilla*) in the Dzanga-Sangha region of southwestern Central African Republic. *Mammalia*, **52**, 309–23.

Carter, J. 1981. Journey to freedom. *Smithsonian*, **12**, 90–101.

Carter, J. 1988. Freed from keepers and cages, chimps come of age on Baboon Island. *Smithsonian*, **19**, 36–49.

Cavallo, J. A. & Blumenschine, R. J. 1989. Tree-stored leopard kills: expanding the hominid scavenging niche. *Journal of Human Evolution*, **18**, 393–9.

Cerling, T. E. & Hay, R. 1986. An isotopic study of paleosol carbonates from Olduvai Gorge. *Quaternary Research*, **25**, 63–78.

Chance, M. R. A. 1960. Kohler's chimpanzees: how did they perform? *Man*, **60**, 130–5.

Chavaillon, J., Chavaillon, N., Hours, F. & Piperno, M. 1979. From Oldowan to the Middle Stone Age at Melka-Kunture (Ethiopia). Understanding cultural changes. *Quaternaria*, **21**, 87–114.

Cheney, D. L. & Seyfarth, R. M. 1990. *How Monkeys See the World*. Chicago: University of Chicago Press.

Chevalier-Skolnikoff, S., Galdikas, B. M. F. & Skolnikoff, A. 1982. The adaptive significance of higher intelligence in wild orang-utans: a preliminary report. *Journal of Human Evolution*, **11**, 639–52.

Clark, J. D. & Harris, J. W. K. 1985. Fire and its roles in early hominid lifeways. *African Archaeological Review*, **3**, 3–27.

Clutton-Brock, T. H. & Gillett, J. B. 1979. A survey of forest composition in the Gombe National Park, Tanzania. *African Journal of Ecology*, **17**, 131–58.

Clutton-Brock, T. H. & Harvey, P. H. 1980. Primates, brains and ecology. *Journal of Zoology*, **190**, 309–23.

Cohen, Y. A. 1968. Culture as adaptation. In: *Man in Adaptation. The Cultural Present*, ed. Y. A. Cohen, pp. 40–60. Chicago: Aldine.

Collins, D. A. & McGrew, W. C. 1985. Chimpanzees' (*Pan troglodytes*) choice of prey among termites (Macrotermitinae) in western Tanzania. *Primates*, **26**, 375–89.

Collins, D. A. & McGrew, W. C. 1987. Termite fauna related to differences in tool-use between groups of chimpanzees (*Pan troglodytes*). *Primates*, **28**, 457–71.

Collins, D. A. & McGrew, W. C. 1988. Habitats of three groups of chimpanzees (*Pan troglodytes*) in western Tanzania compared. *Journal of Human Evolution*, **17**, 553–74.

Conklin, N. L. & Wrangham, R. W. 1991. Dietary strategies and nutrient intakes of chimpanzees in Kibale Forest, Uganda. *Philosophical Transactions of the Royal Society of London, Series B*, **334**, 11–18.

Connolly, K. 1974. The development of skill. *New Scientist*, **62**, 537–40.

Coon, C. S. 1971. *The Hunting Peoples*. Boston: Little, Brown.

Cornevin, R. 1969. *Histoire de Togo*. Paris: Berger-Levrault.

Dahlberg, F. (ed.) 1981. *Woman the Gatherer*. New Haven: Yale University Press.

Dart, R. A. 1949. The predatory implemental technique of *Australopithecus*. *American Journal of Physical Anthropology*, **7**, 1–38.

Dart, R. A. 1956. Cultural status of the South African man-apes. In: *Smithsonian Report for 1955*, pp. 317–38. Washington: Smithsonian Institution.

Davidson, I. & Noble, W. 1989. The archaeology of perception: traces of depiction and language. *Current Anthropology*, **30**, 125–55.

Dennell, R. W., Rendell, H. M. & Hailwood, E. 1988. Late Pliocene artefacts from northern Pakistan. *Current Anthropology*, **29**, 495–8.

Diamond, J. 1987. Bower building and decoration by the bowerbird *Amblyornis inornatus*. *Ethology*, **74**, 177–204.

Diamond, J. 1988. Experimental study of bower decoration by the bowerbird *Amblyornis inornatus*, using colored poker chips. *American Naturalist*, **131**, 631–53.

Diamond, J. 1990. The future of DNA–DNA hybridization studies. *Journal of Molecular Evolution*, **30**, 196–201.

Dienske, H. & van Vreeswijk, W. 1987. Regulation of nursing in chimpanzees. *Developmental Psychobiology*, **20**, 71–83.

Dobzhansky, T. 1972. On the evolutionary uniqueness of man. *Evolutionary Biology*, **6**, 415–30.

Dunbar, R. I. M. 1988. *Primate Social Systems*. London: Croom Helm.

Dunbar, R. I. M. 1989. Ecological modelling in an evolutionary context. *Folia Primatologica*, **53**, 235–46.

Dunnett, S., van Orshoven, J. & Albrecht, H. 1970. Peaceful co-existence between chimpanzee and man in West Africa. *Bijdragen tot de Dierkunde*, **40**, 148–53.

Duviard, D. & Segeren, P. 1974. La colonisation d'un myrmecophyte, le parasolier, par *Crematogaster* spp. (Myrmicinae) en Côte-d'Ivoire forestière. *Insectes Sociaux*, **21**, 191–212.

Eaton, G. 1972. Snowball construction by a feral troop of Japanese macaques (*Macaca fuscata*) living under seminatural conditions. *Primates*, **13**, 411–14.

Eaton, R. L. 1978. Do chimpanzees use weapons or mimic hominid weaponry? *Carnivore*, **1**(2), 82–9.

Eaton, S. B. & Konner, M. 1985. Paleolithic nutrition: a consideration of its nature and current implications. *New England Journal of Medicine*, **312**, 283–9.

Eggeling, W. J. 1947. Observations on the ecology of the Budongo rain forest, Uganda. *Journal of Ecology*, **34**, 20–87.

Eisenberg, J. F. 1973. Mammalian social systems: are primate social systems unique? In: *Precultural Primate Behavior*, ed. E. W. Menzel, pp. 232–49. Basel: Karger.

Etkin, W. 1954. Social behavior and the evolution of man's mental faculties. *American Naturalist*, **88**, 129–42.

Falk, D. 1987. Brain lateralization in primates and its evolution in hominids. *Yearbook of Physical Anthropology*, **30**, 107–25.

Falk, J. L. 1958. The grooming behavior of the chimpanzee as a reinforcer. *Journal of the Experimental Analysis of Behavior*, **1**, 83–5.

Fay, J. M. 1989. Partial completion of a census of the western lowland gorilla ((*Gorilla g. gorilla*) (Savage and Wyman)) in southwestern Central African Republic. *Mammalia*, **53**, 203–15.

Fay, J. M. & Carroll, R. W. 1992. Pounding tools, termite-mound digging sticks, and termite-tunnel probing stalks of chimpanzees (*Pan t. troglodytes*) in the Central African Republic and People's Republic of Congo. Unpublished manuscript.

Feistner, A. T. C. & McGrew, W. C. 1989. Food-sharing in primates: a critical review. In: *Perspectives in Primate Biology*, vol. 3, ed. P. K. Seth and S. Seth, pp. 21–36. New Delhi: Today and Tomorrow's Publishers.

Fischer, G. J. & Kitchener, S. L. 1965. Comparative learning in young gorillas and orang-utans. *Journal of Genetic Psychology*, **107**, 337–48.

Fleming, C. A. 1951. Sea lions as geological agents. *Journal of Sedimentary Petrology*, **21**, 22–5.

Fletcher, D. J. C. 1978. The African bee, *Apis mellifera adansonii*, in Africa. *Annual Review of Entomology*, **23**, 151–72.

Foley, R. 1982. A reconsideration of the role of predation on large mammals in tropical hunter-gatherer adaptation. *Man*, **17**, 393–402.

Foley, R. (ed.) 1984. *Hominid Evolution and Community Ecology*. London: Academic Press.

Foley, R. 1987*a*. *Another Unique Species*. Harlow: Longman.

Foley, R. 1987*b*. Hominid species and stone-tool assemblages: how are they related? *Antiquity*, **61**, 380–92.

Foley, R. A. 1991. How useful is the culture concept in early hominid studies? In: *The Origins of Human Behaviour*, ed. R. A. Foley, pp. 25–38. London: Unwin Hyman.

Foley, R. A. & Lee, P. C. 1989. Finite social space, evolutionary pathways, and reconstructing hominid behavior. *Science*, **243**, 901–6.
Fossey, D. & Harcourt, A. H. 1977. Feeding ecology of free-ranging mountain gorilla (*Gorilla gorilla beringei*). In: *Primate Ecology*, ed. T. H. Clutton-Brock, pp. 415–47. London: Academic Press.
Fouts, D. H. 1989. Signing interactions between mother and infant chimpanzees. In: *Understanding Chimpanzees*, ed. P. G. Heltne and L. A. Marquardt, pp. 242–51. Cambridge, Mass.: Harvard University Press.
Fouts, R., Chown, W. & Goodin, L. 1974. Translation of signed responses in American Sign Language from vocal English stimuli to physical object stimuli by a chimpanzee (*Pan*). *Learning and Motivation*, **7**, 458–75.
Fouts, R. S. & Couch, J. B. 1976. Cultural evolution of learned language in chimpanzees. In: *Communicative Behavior and Evolution*, ed. M. E. Hahn and E. C. Simmel, pp. 141–61. New York: Academic Press.
Frisch, J. E. 1959. Research on primate behavior in Japan. *American Anthropologist*, **61**, 584–96.
Fritz, P. & Fritz, J. 1979. Resocialization of chimpanzees: ten years of experience at the Primate Foundation of Arizona. *Journal of Medical Primatology*, **8**, 202–21.
Fry, G. F. 1985. Analysis of fecal material. In: *The Analysis of Prehistoric Diets*, ed. R. I. Gilbert and J. H. Mielke, pp. 127–54. Orlando: Academic Press.
Galdikas, B. M. F. 1982. Orang-utan tool-use at Tanjung Puting Reserve, Central Indonesian Borneo (Kalimantan Tengah). *Journal of Human Evolution*, **10**, 19–33.
Galdikas, B. M. F. 1989. Orangutan tool use. *Science*, **243**, 152.
Galdikas, B. M. F. & Teleki, G. 1981. Variations in subsistence activities of female and male pongids: new perspectives on the origins of hominid labor division. *Current Anthropology*, **22**, 241–56.
Galef, B. G. 1976. Social transmission of acquired behavior: a discussion of tradition and social learning in vertebrates. *Advances in the Study of Behavior*, **6**, 77–100.
Galef, B. G. 1990. Tradition in animals: field observations and laboratory analyses. In: *Interpretation and Explanation in the Study of Animal Behavior*, ed. M. Bekoff and D. Jamieson, pp. 74–95. Boulder: Westview Press.
Gallup, G. G. 1987. Self-awareness. In: *Comparative Primate Biology*, vol. 2, part B: Behavior, Cognition, and Motivation, ed. G. Mitchell and J. Erwin, pp. 3–16. New York: Alan R. Liss.
Gardner, R. A. & Gardner, B. T. 1969. Teaching sign language to a chimpanzee. *Science*, **165**, 664–72.
Garner, R. L. 1896. *Gorillas and Chimpanzees*. London: Osgood, McIlvaine.
Gartlan, J. S. & Struhsaker, T. T. 1972. Polyspecific associations and niche separation of rain forest anthropoids in Cameroon, West Africa. *Journal of Zoology*, **168**, 221–66.
Geerling, C. & Bokdam, J. 1973. Fauna of the Comoe National Park, Ivory Coast. *Biological Conservation*, **5**, 251–7.
Ghiglieri, M. P. 1984. *The Chimpanzees of Kibale Forest*. New York: Columbia University Press.
Ghiglieri, M. P. 1987. Sociobiology of the great apes and the hominid ancestor. *Journal of Human Evolution*, **16**, 319–57.

Ghiglieri, M. P. 1988. *East of the Mountains of the Moon*. New York: Free Press.
Gibson, K. R. 1991. Tools, language and intelligence: evolutionary implications. *Man*, **26**, 255–64.
Gomez, J. C. 1988. Tool-use and communication as alternative strategies of problem solving in the gorilla. *Primate Report*, **19**, 25–8.
Goodall, A. 1979. *The Wandering Gorillas*. London: Collins.
Goodall, J. 1962. Nest building behavior in the free ranging chimpanzee. *Annals of the New York Academy of Sciences*, **102**, 455–67.
Goodall, J. 1963. Feeding behaviour of wild chimpanzees. *Symposia of the Zoological Society of London*, **10**, 39–48.
Goodall, J. 1964. Tool-using and aimed throwing in a community of free-living chimpanzees. *Nature*, **201**, 1264–6.
Goodall, J. v. L. 1967. *My Friends the Wild Chimpanzees*. Washington: National Geographic Society.
Goodall, J. v. L. 1968. The behavior of free-living chimpanzees in the Gombe Stream Reserve. *Animal Behaviour Monographs*, **1**, 161–311.
Goodall, J. v. L. 1970. Tool-using in primates and other vertebrates. *Advances in the Study of Behavior*, **3**, 195–249.
Goodall, J. v. L. 1971. *In the Shadow of Man*. London: Collins.
Goodall, J. v. L. 1973. Cultural elements in a chimpanzee community. In: *Precultural Primate Behavior*, ed. E. W. Menzel, pp. 144–84. Basel: S. Karger.
Goodall, J. 1986. *The Chimpanzees of Gombe*. Cambridge, Mass.: Harvard University Press.
Goodall, J. 1990. *Through a Window*. London: Weidenfeld and Nicolson.
Goodall, J. & Hamburg, D. A. 1974. Chimpanzee behavior as a model for the behavior of early man: new evidence on possible origins of human behavior. *American Handbook of Psychiatry*, **6**, 14–43.
Goodall, J. v. L. & van Lawick, H. 1966. On the use of tools by the Egyptian vulture, *Neophron percnopterus*. *Nature*, **212**, 1468–9.
Goodman, M., Tagle, D. A., Fitch, D. H. A., Bailey, W., Czelsniak, J., Koop, B. F., Benson, P. & Slighton, J. L. 1990. Primate evolution at the DNA level and a classification of hominoids. *Journal of Molecular Evolution*, **30**, 260–6.
Gotwald, W. H. 1974. Predatory behavior and food preferences of driver ants in selected African habitats. *Annals of the Entomological Society of America*, **67**, 877–86.
Goudsblom, J. 1986. The human monopoly on the use of fire: its origins and conditions. *Human Evolution*, **1**, 517–23.
Gowlett, J. A. J., Harris, J. W. K., Walton, D. & Wood, B. A. 1981. Early archaeological sites, hominid remains, and traces of fire from Chesowanja, Kenya. *Nature*, **294**, 125–9.
Green, S. 1975. Dialects in Japanese monkeys: vocal learning and cultural transmission of locale-specific behaviour? *Zeitschrift für Tierpsychologie*, **38**, 301–14.
Griffiths, J. F. (ed.) 1972. *Climates of Africa*. Amsterdam: Elsevier.
Grine, F. E. & Kay, R. F. 1988. Early hominid diets from quantitative image analysis of dental microwear. *Nature*, **333**, 765–8.
Groves, C. P. 1986. Systematics of the great apes. In: *Comparative Primate Biology*, vol. 1, Systematics, Evolution, and Anatomy, ed. G. Mitchell and J. Erwin, pp. 187–217. New York: Alan R. Liss.

Groves, C. P. & Humphrey, N. K. 1973. Asymmetry in gorilla skulls: evidence of lateralized brain function? *Nature*, **244**, 53–4.
Groves, C. P. & Sabater Pí, J. 1985. From ape's nest to human fix-point. *Man*, **20**, 22–47.
Haddow, A. J. 1958. Chimpanzees. *Uganda Wild Life and Sport*, **1**(3), 18–20.
Hall, E. T. 1959. *The Silent Language*. Greenwich: Fawcett.
Hall, K. R. L. & Schaller, G. B. 1964. Tool-using behavior of the Californian sea otter. *Journal of Mammalogy*, **45**, 287–98.
Hallowell, A. I. 1960. Self, society, and culture in phylogenetic perspective. In: *The Evolution of Man*, vol. 2, ed. S. Tax, pp. 309–72. Chicago: University of Chicago Press.
Hamburg, D. A. & McCown, E. R. (eds.) 1979. *The Great Apes*. Menlo Park: Benjamin/Cummings.
Hamilton, W. J., Buskirk, R. E. & Buskirk, W. H. 1975. Defensive stoning by baboons. *Nature*, **256**, 488–9.
Hannah, A. C. 1989. Rehabilitation of captive chimpanzees (*Pan troglodytes verus*). PhD thesis, University of Stirling.
Hannah, A. C. & McGrew, W. C. 1987. Chimpanzees using stones to crack open oil palm nuts in Liberia. *Primates*, **28**, 31–46.
Hannah, A. C. & McGrew, W. C. 1991. Rehabilitation of captive chimpanzees. In: *Primate Responses to Environmental Change*, ed. H. O. Box, pp. 167–86. London: Chapman and Hall.
Hansell, M. 1987. What's so special about using tools? *New Scientist*, **113**, 54–6.
Harcourt, A. H. & Harcourt, S. A. 1984. Insectivory by gorillas. *Folia Primatologica*, **43**, 229–33.
Hardesty, D. L. 1972. The human ecological niche. *American Anthropologist*, **74**, 458–66.
Harding, R. S. O. 1984. Primates of the Kilimi area, northwest Sierra Leone. *Folia Primatologica*, **42**, 96–114.
Harding, R. S. O. & Strum, S. C. 1976. The predatory baboons of Kekopey. *Natural History*, [March], 49–53.
Harris, J. W. K. 1983. Cultural beginnings: Plio-Pleistocene archaeological occurrences from the Afar, Ethiopia. *African Archaeological Review*, **1**, 3–31.
Harris, M. 1964. *The Nature of Cultural Things*. New York: Random House.
Harris, M. 1979. *Cultural Materialism*. New York: Vintage.
Harris, M. 1985. *The Sacred Cow and the Abominable Pig*. New York: Touchstone.
Hart, H. & Panzer, A. 1925. Have subhuman animals culture? *American Journal of Sociology*, **30**, 703–9.
Hart, T. B. & Hart, J. A. 1986. The ecological basis of hunter-gatherer subsistence in African rain forests: the Mbuti of eastern Zaïre. *Human Ecology*, **14**, 29–55.
Hartley, C. W. S. 1966. *The Oil Palm (Elaeis guineensis Jacq.)*. London: Longmans, Green.
Hasegawa, T., Hiraiwa, M., Nishida, T. & Takasaki, H. 1983. New evidence on scavenging behavior in wild chimpanzees. *Current Anthropology*, **24**, 231–2.
Hauser, M. D. & Wrangham, R. W. 1987. Manipulation of food calls in captive chimpanzees: a preliminary report. *Folia Primatologica*, **48**, 207–10.

Hausfater, G. 1975. Predatory behavior of yellow baboons. *Behavior*, **56**, 44–68.

Hayes, C. 1951. *The Ape in Our House*. New York: Harper.

Hayes, K. J. & Hayes, C. 1952. Imitation in a home-raised chimpanzee. *Journal of Comparative and Physiological Psychology*, **45**, 450–9.

Hayes, K. J. & Hayes, C. 1954. The cultural capacity of chimpanzee. *Human Biology*, **26**, 288–303.

Haynes, V. 1973. The Calico site: artifacts or geofacts? *Science*, **181**, 305–10.

Headland, T. N. & Reid, L. A. 1989. Hunter-gatherers and their neighbors from prehistory to the present. *Current Anthropology*, **30**, 43–66.

Hediger, H. 1977. Nest and home. *Folia Primatologica*, **28**, 170–87.

Henty, C. J. 1986. Development of snail-smashing in song thrushes. *British Birds*, **79**, 277–81.

Hewes, G. W. 1961. Food transport and the origin of hominid bipedalism. *American Anthropologist*, **63**, 687–710.

Hiatt, B. 1967. The food quest and the economy of the Tasmanian aborigines. *Oceania*, **38**, 99–138.

Hiatt, B. 1968. The food quest and the economy of the Tasmanian aborigines [continued]. *Oceania*, **38**, 190–219.

Hill, K. 1982. Hunting and human evolution. *Journal of Human Evolution*, **11**, 521–44.

Hill, W. C. O. 1963. The Ufiti: the present position. *Symposia of the Zoological Society of London*, **10**, 57–9.

Hill, W. C. O. 1969. The discovery of the chimpanzee. In: *The Chimpanzee*, vol. 1, pp. 1–21. Basel: Karger.

Hinde, R. A. & Fisher, J. 1952. Further observations on the opening of milk bottles by birds. *British Birds*, **44**, 393–6.

Hiraiwa-Hasegawa, M. 1989. Sex differences in the behavioral development of chimpanzees at Mahale. In: *Understanding Chimpanzees*, ed. P. G. Heltne and L. A. Marquardt, pp. 104–15. Cambridge, Mass.: Harvard University Press.

Hladik, C. M. 1973. Alimentation et activité d'un groupe de chimpanzés réintroduits en forêt Gabonaise. *La Terre et la Vie*, **27**, 343–413.

Hladik, C. M. 1977. Chimpanzees of Gabon and chimpanzees of Gombe: some comparative data on the diet. In: *Primate Ecology*, ed. T. H. Clutton-Brock, pp. 481–501. London: Academic Press.

Hladik, C. M. & Viroben, G. 1974. L'alimentation protéique du chimpanzé dans son environnement forestier naturel. *Comptes Rendus, Serie D*, **279**, 1475–8.

Hodder, I. 1977. The distribution of material culture items in the Baringo District, Western Kenya. *Man*, **12**, 239–69.

Holldobler, B. & Wilson, E. O. 1977. Weaver ants. *Scientific American*, **237**(6), 146–54.

Holloway, R. L. 1969. Culture: a *human* domain. *Current Anthropology*, **10**, 395–412.

Holloway, R. L. & de la Coste-Lareymondie, M. C. 1982. Brain endocast asymmetry in pongids and hominids: some preliminary findings on the paleontology of cerebral dominance. *American Journal of Physical Anthropology*, **58**, 101–10.

van Hooff, J. A. R. A. M. 1970. A component analysis of the structure of social behaviour of a semi-captive chimpanzee group. *Experientia*, **26**, 549–50.

van Hooff, J. A. R. A. M. 1973. The Arnhem Zoo chimpanzee consortium: an attempt to create an ecologically and socially acceptable habitat. *International Zoo Yearbook*, **13**, 195–205.

Horn, A. D. 1980. Some observations on the ecology of the bonobo chimpanzee (*Pan paniscus*, Schwarz 1929) near Lake Tumba, Zaïre. *Folia Primatologica*, **34**, 145–69.

Howse, P. E. 1970. *Termites*. London: Hutchinson.

Hrdy, S. B. 1981. *The Woman That Never Evolved*. Cambridge, Mass.: Harvard University Press.

Huffman, M. A. 1984. Stone-play of *Macaca fuscata* in Arahiyama B troop: transmission of a non-adaptive behavior. *Journal of Human Evolution*, **13**, 725–35.

Huffman, M. A. & Quiatt, D. 1986. Stone handling by Japanese macaques (*Macaca fuscata*): implications for tool use of stone. *Primates*, **27**, 413–23.

Huffman, M. A. & Seifu, M. 1989. Observations on the illness and consumption of a possibly medicinal plant *Vernonia amygdalina* (Del.) by a wild chimpanzee in the Mahale Mountains National Park, Tanzania. *Primates*, **30**, 51–63.

Huntingford, G. W. B. 1955. The economic life of the Dorobo. *Anthropos*, **50**, 602–34.

Ichikawa, M. 1981. Ecological and sociological importance of honey to the Mbuti net hunters, eastern Zaïre. *African Study Monographs*, **1**, 55–68.

Ichikawa, M. 1983. An examination of the hunting-dependent life of the Mbuti pygmies, eastern Zaïre. *African Study Monographs*, **4**, 55–76.

Ichikawa, M. 1987. Food restrictions of the Mbuti pygmies, eastern Zaïre. *African Study Monographs* (Supplementary Issue), **6**, 97–121.

Imanishi, K. 1957. Social behaviour of Japanese monkeys, *Macaca fuscata*. *Psychologia*, **1**, 47–54.

Ingold, T. 1986*a*. Extraction, appropriation and co-operation: the constituents of human hunting. In: *The Appropriation of Nature*, ed. T. Ingold, pp. 101–29. Manchester: Manchester University Press.

Ingold, T. 1986*b*. Gatherer-hunter, forager-predator: modes of subsistence in human evolution. In: *The Appropriation of Nature*, ed. T. Ingold, pp. 79–100. Manchester: Manchester University Press.

Ingold, T. 1986*c*. Prologue: concerning the hunter, and his spear. In: *The Appropriation of Nature*, ed. T. Ingold, pp. 1–15. Manchester: Manchester University Press.

Ingold, T. 1986*d*. The significance of storage in hunting societies. In: *The Appropriation of Nature*, ed. T. Ingold, pp. 198–221. Manchester: Manchester University Press.

Isaac, B. 1987. Throwing and human evolution. *African Archaeological Review*, **5**, 3–17.

Isaac, G. L. 1976. Stages of cultural elaboration in the Pleistocene: possible archaeological indicators of the development of language capacities. *Annals of the New York Academy of Sciences*, **280**, 275–88.

Isaac, G. L. 1978*a*. Food sharing and human evolution: archaeological evidence from the Plio-Pleistocene of East Africa. *Journal of Anthropological Research*, **34**, 311–25.

Isaac, G. L. 1978*b*. The food-sharing behavior of protohuman hominids. *Scientific American*, **238**(4), 90–108.

Isaac, G. L. 1981. Archaeological tests of alternative models of early hominid

behaviour: excavation and experiments. *Philosophical Transactions of the Royal Society of London, Series B*, **292**, 177–88.
Isabirye-Basuta, G. 1988. Food competition among individuals in a free-ranging chimpanzee community in Kibale Forest, Uganda. *Behaviour*, **105**, 135–47.
Isabirye-Basuta, G. 1989. Feeding ecology of chimpanzees in the Kibale Forest, Uganda. In: *Understanding Chimpanzees*, ed. P. G. Heltne and L. A. Marquardt, pp. 116–27. Cambridge, Mass.: Harvard University Press.
Itani, J. 1979. Distribution and adaptation of chimpanzees in an arid area. In: *The Great Apes*, ed. D. A. Hamburg and E. R. McCown, pp. 55–71. Menlo Park: Benjamin/Cummings.
Itani, J. & Nishimura, A. 1973. The study of infrahuman culture in Japan. In: *Precultural Primate Behavior*, ed. E. W. Menzel, pp. 26–50. Basel: S. Karger.
Itani, J. & Suzuki, A. 1967. The social unit of chimpanzees. *Primates*, **8**, 355–81.
Izawa, K. 1970. Unit groups of chimpanzees and their nomadism in the savanna woodland. *Primates*, **11**, 1–46.
Izawa, K. & Itani, J. 1966. Chimpanzees of the Kasakati basin, Tanganyika. I. Ecological study in the rainy season 1963–1964. *Kyoto University African Studies*, **1**, 73–156.
Izawa, K. & Mizuno, A. 1977. Palm-fruit cracking behaviour of wild black-capped capuchin (*Cebus apella*). *Primates*, **18**, 773–93.
James, H. W. & Brooke, R. K. 1971. Stone-built base of the nest of the grey-backed finchlark. *Ostrich*, **42**, 226.
James, S. R. 1989. Hominid use of fire in the lower and middle Pleistocene. *Current Anthropology*, **30**, 1–26.
Janzen, D. H. 1977. Why fruits rot, seeds mold, and meat spoils. *American Naturalist*, **111**, 691–713.
Janzen, D. H. 1979. How to be a fig. *Annual Review of Ecology and Systematics*, **10**, 13–52.
Janzen, D. H. & Schoener, T. W. 1968. Differences in insect abundance and diversity between wetter and drier sites during a tropical dry season. *Ecology*, **49**, 96–110.
Jeffrey, S. 1975. Ghana's new forest national park. *Oryx*, **13**, 34–6.
Jerison, H. J. 1973. *Evolution of the Brain and Intelligence*. New York: Academic Press.
Jolly, A. 1991. Conscious chimpanzees? A review of recent literature. In: *Cognitive Ethology: The Minds of Other Animals, Essays in Honor of Donald R. Griffin*, ed. C. A. Ristau, pp. 231–52. Hillsdale: Lawrence Erlbaum Associates.
Jolly, C. J. 1970. The seed-eaters: a new model of hominid differentiation based on a baboon analogy. *Man*, **5**, 5–26.
Jones, C. & Sabater Pí, J. 1969. Sticks used by chimpanzees in Rio Muni, West Africa. *Nature*, **223**, 100–1.
Jones, C. & Sabater Pí, J. 1971. Comparative ecology of *Gorilla gorilla* (Savage and Wyman) and *Pan troglodytes* (Blumenbach) in Rio Muni, West Africa. *Bibliotheca Primatologica*, **13**, 1–96.
Jones, R. 1977. The Tasmanian paradox. In: *Stone Tools as Cultural Markers: Change, Evolution and Complexity*, ed. R. V. S. Wright, pp. 189–204. Canberra: Australian Institute of Aboriginal Studies.

Jones, R. 1984. Hunters and history: a case study from western Tasmania. In: *Past and Present in Hunter Gatherer Studies*, ed. C. Schrire, pp. 27–65. Orlando: Academic Press.

Jordan, C. 1982. Object manipulation and tool-use in captive pygmy chimpanzees (*Pan paniscus*). *Journal of Human Evolution*, **11**, 35–9.

Kalb, J. E., Jolly, C. J., Oswald, E. B. & Whitehead, P. F. 1984. Early hominid habitation in Ethiopia. *American Scientist*, **72**, 168–78.

Kano, T. 1971. The chimpanzees of Filabanga, western Tanzania. *Primates*, **12**, 229–46.

Kano, T. 1972. Distribution and adaptation of the chimpanzee on the eastern shore of Lake Tanganyika. *Kyoto University African Studies*, **7**, 37–129.

Kano, T. 1979. A pilot study on the ecology of pygmy chimpanzees, *Pan paniscus*. In: *The Great Apes*, ed. D. A. Hamburg and E. R. McCown, pp. 123–35. Menlo Park: Benjamin/Cummings.

Kano, T. 1982. The use of leafy twigs for rain cover by the pygmy chimpanzees of Wamba. *Primates*, **23**, 453–7.

Kano, T. 1983. An ecological study of the pygmy chimpanzees (*Pan paniscus*) of Yalosidi, Republic of Zaïre. *International Journal of Primatology*, **4**, 1–31.

Kano, T. 1984. Distribution of pygmy chimpanzees (*Pan paniscus*) in the Central Zaïre Basin. *Folia Primatologica*, **43**, 36–52.

Kano, T. & Mulavwa, M. 1984. Feeding ecology of the pygmy chimpanzee (*Pan paniscus*) of Wamba. In: *The Pygmy Chimpanzee*, ed. R. L. Susman, pp. 233–74. New York: Plenum Press.

Kawabe, M. 1966. One observed case of hunting behavior among wild chimpanzees living in the savanna woodland of western Tanzania. *Primates*, **7**, 393–6.

Kawamura, S. 1959. The process of sub-culture propagation among Japanese macaques. *Primates*, **2**, 43–60.

Kawamura, S. 1972. The pre-culture in the Japanese macaques. In: *Modèles Animaux du Comportement Humain*, pp. 155–79. Paris: Centre National de Recherche Scientifique.

Kawanaka, K. 1982. Further studies on predation by chimpanzees of the Mahali Mountains. *Primates*, **23**, 364–84.

Kellogg, W. N. 1969. Research on the home-raised chimpanzee. *The Chimpanzee*, **1**, 369–92.

Kellogg, W. N. & Kellogg, L. A. 1933. *The Ape and the Child*. New York: McGraw-Hill.

Kiernan, K., Jones, R. & Ranson, D. 1983. New evidence from Fraser Cave for glacial man in south west Tasmania. *Nature*, **301**, 28–32.

King, B. J. 1991. Social information transfer in monkeys, apes, and hominids. *Yearbook of Physical Anthropology*, **34**, 97–115.

Kitahara-Frisch, J. 1977. Tools or toys? What have we really learned from wild chimpanzees about tool use? *Journal of the Anthropological Society of Nippon*, **85**, 57–64.

Kitahara-Frisch, J. & Norikoshi, K. 1982. Spontaneous sponge-making in captive chimpanzees. *Journal of Human Evolution*, **11**, 41–7.

Kitahara-Frisch, J. Norikoshi, K. & Hara, K. 1987. Use of a bone fragment as a step towards secondary tool use in captive chimpanzees. *Primate Report*, **18**, 33–7.

Kock, D. 1967. Die Verbreitung des Schimpansen, *Pan troglodytes schweinfurthii* (Giglioli, 1872) im Sudan. *Zeitschrift für Säugetierkunde*, **32**, 250–5.

Köhler, W. 1927. *The Mentality of Apes*, 2nd edn. London: Kegan Paul, Trench, Trubner.

Kohts, N. 1935. *Infant Ape and Human Child (Instincts, Emotions, Play, Habits)*. Moscow: Scientific Memoirs of the Museum Darwinianum.

Kollar, E. J. 1972. Object relations and the origin of tools. *Archives of General Psychiatry*, **26**, 23–7.

Kortlandt, A. 1962. Chimpanzees in the wild. *Scientific American*, **206**, 128–38.

Kortlandt, A. 1965. How do chimpanzees use weapons when fighting leopards? *Year Book of the American Philosophical Society*, [1965], 327–32.

Kortlandt, A. 1967. Experimentation with chimpanzees in the wild. In: *Neue Ergebnisse der Primatologie*, ed. D. Starck, R. Schneider and H.-J. Kuhn, pp. 208–24. Stuttgart: Gustav Fischer.

Kortlandt, A. 1980. How might early hominids have defended themselves against predators and food competitors? *Journal of Human Evolution*, **9**, 79–112.

Kortlandt, A. 1983. Marginal habitats of chimpanzees. *Journal of Human Evolution*, **12**, 231–78.

Kortlandt, A. 1986. The use of stone tools by wild-living chimpanzees and earliest hominids. *Journal of Human Evolution*, **15**, 77–132.

Kortlandt, A. & Holzhaus, E. 1987. New data on the use of stone tools by chimpanzees in Guinea and Liberia. *Primates*, **28**, 473–96.

Kortlandt, A. & Kooij, M. 1963. Protohominid behaviour in primates (preliminary communication). *Symposia of the Zoological Society of London*, **10**, 61–88.

Kraemer, H. C. 1979. A study of reliability and its hierarchical structure in observed chimpanzee behavior. *Primates*, **20**, 553–61.

Krebs, J. R. 1973. Behavior aspects of predation. In: *Perspectives in Ethology*, ed. P. P. G. Bateson and P. H. Klopfer, pp. 73–111. New York: Plenum Press.

Kroeber, A. L. 1928. Sub-human cultural beginnings. *Quarterly Review of Biology*, **3**, 325–42.

Kroeber, A. L. & Kluckhohn, C. 1952. Culture: a critical review of concepts and definitions. *Papers of the Peabody Museum of American Archeology and Ethnology*, **47**, 41–72.

Kummer, H. 1971. *Primate Societies*. Chicago: Aldine-Atherton.

Kummer, H. & Goodall, J. 1985. Conditions of innovative behaviour in primates. *Philosophical Transactions of the Royal Society of London, Series B*, **308**, 203–14.

Kuroda, S. 1980. Social behavior of the pygmy chimpanzee. *Primates*, **21**, 181–97.

Kuroda, S. 1984. Interaction over food among pygmy chimpanzees. In: *The Pygmy Chimpanzee*, ed. R. L. Susman, pp. 301–24. New York: Plenum Press.

Lane, H. 1977. *The Wild Boy of Aveyron*. London: Granada.

Laughlin, W. S. 1968. Hunting: an integrating biobehavior system and its evolutionary importance. In: *Man the Hunter*, ed. R. B. Lee and I. DeVore, pp. 304–20. Chicago: Aldine-Atherton.

Leakey, L. S. B. 1968. Bone-smashing by late Miocene Hominidae. *Nature*, **218**, 528–30.
Leakey, M. D. 1966. A review of the Oldowan culture from Olduvai Gorge, Tanzania. *Nature*, **210**, 462–6.
Leakey, M. D. 1971. *Olduvai Gorge*, vol. 3. Cambridge: Cambridge University Press.
Leakey, M. D. 1975. Cultural patterns in the Olduvai sequence. In: *After the Australopithecines*, ed. K. W. Butzer and G. L. Isaac, pp. 447–93. The Hague: Mouton.
Lee, P. C., Thornback, J. & Bennett, E. L. 1988. *Threatened Primates of Africa. The IUCN Red Data Book*. Gland: International Union for the Conservation of Nature and Natural Resources.
Lee, R. B. 1968. What hunters do for a living, or, How to make out on scarce resources. In: *Man the Hunter*, ed. R. B. Lee and I. DeVore, pp. 30–48. Chicago: Aldine-Atherton.
Lee, R. B. 1979. *The !Kung San*. Cambridge: Cambridge University Press.
Lee, R. B. & DeVore, I. (eds.) 1968. *Man the Hunter*. Chicago: Aldine-Atherton.
Lee, R. B. & DeVore, I. (eds.) 1976. *Kalahari Hunter-Gatherers*. Cambridge, Mass.: Harvard University Press.
Le Gros Clark, W. E. 1967. Human food habits as determining the basic patterns of economic and social life. In: *Proceedings of the Seventh International Congress of Nutrition*, vol. 4, ed. J. Kuhnau, pp. 18–24. Braunschweig: Vieweg und Sohn.
Lemay, M. 1976. Morphological cerebral asymmetries of modern man, fossil man, and nonhuman primates. *Annals of the New York Academy of Sciences*, **280**, 349–66.
Lemonnier, P. 1986. The study of material culture today: towards an anthropology of technical systems. *Journal of Anthropological Archaeology*, **5**, 147–86.
Leroi-Gourhan, A. 1975. The flowers found with Shanidar IV, a Neanderthal burial in Iraq. *Science*, **190**, 562–3.
Lethmate, J. 1982. Tool-using skills of orang-utans. *Journal of Human Evolution*, **11**, 49–64.
Lethmate, J. 1991. Haben Schimpansen eine materielle Kultur? *Biologie in unserer Zeit*, **21**, 132–9.
Lethmate, J. & Dücker, G. 1973. Untersuchungen zum Selbsterkennen im Spiegel bei Orang-utans und einigen anderen Affenarten. *Zeitschrift für Tierpsychologie*, **33**, 248–69.
Lewin, R. 1988. New views emerge on hunters and gatherers. *Science*, **240**, 1146–8.
Longhurst, C., Johnson, R. A. & Wood, T. G. 1978. Predation by *Megaponera foetens* (Fabr.) (Hymenoptera: Formicidae) on termites in the Nigerian southern Guinea savanna. *Oecologia*, **32**, 101–7.
Lovejoy, C. O. 1981. The origin of man. *Science*, **211**, 341–50.
Lustig-Arecco, V. 1975. *Technology*. New York: Holt, Rinehart and Winston.
Machlis, L., Dodd, P. W. D. & Fentress, J. C. 1985. The pooling fallacy: problems arising when individuals contribute more than one observation to the data set. *Zeitschrift für Tierpsychologie*, **68**, 201–14.

MacKinnon, J. 1974. The behaviour and ecology of wild orang-utans (*Pongo pygmaeus*). *Animal Behaviour*, **22**, 3–74.

MacNeilage, P. F., Studdert-Kennedy, M. G. & Lindblom, B. 1987. Primate handedness reconsidered. *Behavioral and Brain Sciences*, **10**, 247–303.

Maki, S., Alford, P. L., Bloomsmith, M. A. & Franklin, J. 1989. Food puzzle device simulating termite fishing for captive chimpanzees (*Pan troglodytes*). *American Journal of Primatology, Supplement*, **1**, 71–8.

Malenky, R. K. & Stiles, E. W. 1991. Distribution of terrestrial herbaceous vegetation and its consumption by *Pan paniscus* in the Lomako Forest, Zaïre. *American Journal of Primatology*, **23**, 153–69.

Mann, A. 1972. Hominid and cultural origins. *Man*, **7**, 379–86.

Maples, W. R. 1969. Adaptive behavior of baboons. *American Journal of Physical Anthropology*, **31**, 107–9.

Maples, W. R., Maples, M. K., Greenhood, W. F. & Walek, M. L. 1976. Adaptations of crop-raiding baboons in Kenya. *American Journal of Physical Anthropology*, **45**, 309–16.

Marchant, L. F. 1992. Do chimpanzees show hand preference in spontaneous behavior? *American Journal of Primatology* (in press).

Marchant, L. F. & McGrew, W. C. 1991. Laterality of function in apes: a meta-analysis of methods. *Journal of Human Evolution*, **21**, 425–38.

Marean, C. W. 1989. Sabertooth cats and their relevance for early hominid diet and evolution. *Journal of Human Evolution*, **18**, 559–82.

Markowitz, H. & Spinelli, J. S. 1986. Experimental engineering for primates. In: *Primates: The Road to Self-Sustaining Populations*, ed. K. Benirschke, pp. 489–98. New York: Springer.

Marler, P. & Hobbett, L. 1975. Individuality in a long-range vocalization of wild chimpanzees. *Zeitschrift für Tierpsychologie*, **38**, 97–109.

Marriott, B. & Salzen, E. A. 1979. Food-storing behavior in captive squirrel monkeys (*Saimiri sciureus*). *Primates*, **20**, 307–11.

Martin, P. & Bateson, P. 1986. *Measuring Behaviour*. Cambridge: Cambridge University Press.

Marzke, M. W. & Shackley, M. S. 1986. Hominid hand use in the Pliocene and Pleistocene: evidence from experimental archaeology and comparative morphology. *Journal of Human Evolution*, **15**, 439–60.

Matsuzawa, T. 1985. Colour naming and classification in a chimpanzee (*Pan troglodytes*). *Journal of Human Evolution*, **14**, 283–91.

McBeath, N. M. & McGrew, W. C. 1982. Tools used by wild chimpanzees to obtain termites at Mt. Assirik, Senegal: the influence of habitat. *Journal of Human Evolution*, **11**, 65–72.

McGinnis, P. R. 1979. Sexual behavior of free-living chimpanzees: consort relationships. In: *The Great Apes*, ed. D. A. Hamburg and E. R. McCown, pp. 429–39. Menlo Park: Benjamin/Cummings.

McGrew, W. C. 1974. Tool use by wild chimpanzees in feeding upon driver ants. *Journal of Human Evolution*, **3**, 501–8.

McGrew, W. C. 1975. Patterns of plant food sharing by wild chimpanzees. In: *Contemporary Primatology*, ed. S. Kondo, M. Kawai and A. Ehara, pp. 304–9. Basel: S. Karger.

McGrew, W. C. 1977. Socialization and object manipulation of wild chimpanzees. In: *Primate Bio-Social Development*, ed. S. Chevalier-Skolnikoff and F. E. Poirier, pp. 261–88. New York: Garland.

McGrew, W. C. 1979. Evolutionary implications of sex differences in chimpanzee predation and tool use. In: *The Great Apes*, ed. D. A.

Hamburg and E. R. McCown, pp. 440–63. Menlo Park: Benjamin/Cummings.

McGrew, W. C. 1981*a*. The female chimpanzee as a human evolutionary prototype. In: *Woman the Gatherer*, ed. F. Dahlberg, pp. 35–73. New Haven: Yale University Press.

McGrew, W. C. 1981*b*. Social and cognitive capabilities of non-human primates: lessons from the field to captivity. *International Journal for the Study of Animal Problems*, **2**, 138–49.

McGrew, W. C. 1983. Animal foods in the diets of wild chimpanzees: why cross-cultural variation? *Journal of Ethology*, **1**, 46–61.

McGrew, W. C. 1984. Comment. *Current Anthropology*, **25**, 160.

McGrew, W. C. 1985. The chimpanzee and the oil palm: patterns of culture. *Social Biology and Human Affairs*, **50**, 7–23.

McGrew, W. C. 1987. Tools to get food: the subsistants of Tasmanian aboriginees and Tanzanian chimpanzees compared. *Journal of Anthropological Research*, **43**, 247–58.

McGrew, W. C. 1989*a*. Comment. *Current Anthropology*, **30**, 16–17.

McGrew, W. C. 1989*b*. Recent research on chimpanzees in West Africa. In: *Understanding Chimpanzees*, ed. P. G. Heltne and L. A. Marquardt, pp. 128–33. Cambridge, Mass.: Harvard University Press.

McGrew, W. C. 1989*c*. Why is ape tool use so confusing? In: *Comparative Socioecology*, ed. V. Standen and R. A. Foley, pp. 457–72. Oxford: Blackwell Scientific.

McGrew, W. C. 1991. Chimpanzee material culture: what are its limits and why? In: *The Origins of Human Behaviour*, ed. R. A. Foley, pp. 13–24. London: Unwin Hyman.

McGrew, W. C. 1992*a*. Le cerveau, la main et la pensée: manqué de congruence troublant dans l'usage d'outils chez les grands singes. In: *The Use of Tools in Primates*, ed. J. Chavaillon. Paris: Fondation Fyssen.

McGrew, W. C. 1992*b*. The intelligent use of tools: twenty propositions. In: *Tools, Language and Intelligence: Evolutionary Implications*, ed. T. Ingold and K. R. Gibson. Cambridge: Cambridge University Press.

McGrew, W. C., Baldwin, P. J. & Tutin, C. E. G. 1981. Chimpanzees in a hot, dry and open habitat: Mt. Assirik, Senegal, West Africa. *Journal of Human Evolution*, **10**, 227–44.

McGrew, W. C., Baldwin, P. J. & Tutin, C. E. G. 1988. Diet of wild chimpanzees (*Pan troglodytes verus*) at Mt. Assirik, Senegal. I. Composition. *American Journal of Primatology*, **16**, 213–26.

McGrew, W. C. & Collins, D. A. 1985. Tool-use by wild chimpanzees (*Pan troglodytes*) to obtain termites (*Macrotermes herus*) in the Mahale Mountains, Tanzania. *American Journal of Primatology*, **9**, 47–62.

McGrew, W. C. & Feistner, A. T. C. 1992. Two nonhuman primate models for the evolution of human food-sharing: chimpanzees and callitrichids. In: *The Adapted Mind*, ed. J. W. Barkow, L. Cosmides and J. Tooby. Oxford: Oxford University Press.

McGrew, W. C., Ham, R., Tutin, C. E. G., Fernandez, M. & White, L. 1992. Why don't chimpanzees in Gabon crack nuts? Unpublished manuscript.

McGrew, W. C. & Marchant, L. F. 1992. Chimpanzees, tools, and termites: hand preference or handedness? *Current Anthropology*, **32**, 114–19.

McGrew, W. C. & Rogers, M. E. 1983. Chimpanzees, tools and termites: new records from Gabon. *American Journal of Primatology*, **5**, 171–4.

McGrew, W. C., Sharman, M. J., Baldwin, P. J. & Tutin, C. E. G. 1982. On early hominid plant-food niches. *Current Anthropology*, **23**, 213–14.

McGrew, W. C. & Tutin, C. E. G. 1972. Chimpanzee dentistry. *Journal of the American Dental Association*, **85**, 1198–204.

McGrew, W. C. & Tutin, C. E. G. 1973. Chimpanzee tool use in dental grooming. *Nature*, **241**, 477–8.

McGrew, W. C. & Tutin, C. E. G. 1978. Evidence for a social custom in wild chimpanzees? *Man*, **13**, 234–51.

McGrew, W. C., Tutin, C. E. G. & Baldwin, P. J. 1979*a*. Chimpanzees, tools and termites: cross-cultural comparisons of Senegal, Tanzania, and Rio Muni. *Man*, **14**, 185–214.

McGrew, W. C., Tutin, C. E. G. & Baldwin, P. J. 1979*b*. New data on meat-eating by wild chimpanzees. *Current Anthropology*, **20**, 238–9.

McGrew, W. C., Tutin, C. E. G. & Midgett, P. M. 1975. Tool use in a group of captive chimpanzees. I. Escape. *Zeitschrift für Tierpsychologie*, **37**, 145–62.

Meador, D. M., Rumbaugh, D. M., Pate, J. L. & Bard, K. A. 1987. Learning, problem solving, cognition, and intelligence. In: *Comparative Primate Biology*, vol. 2B, *Behavior, Cognition, and Motivation*, ed. G. Mitchell and J. Erwin, pp. 17–83. New York: Alan R. Liss.

Menzel, E. W. 1972. Spontaneous invention of ladders in a group of young chimpanzees. *Folia Primatologica*, **17**, 87–106.

Menzel, E. W. 1973. Further observations on the use of ladders in a group of young chimpanzees. *Folia Primatologica*, **19**, 450–7.

Menzel, E. W. 1974. A group of young chimpanzees in a one-acre field. *Behavior of Nonhuman Primates*, **5**, 83–153.

Menzel, E. W., Davenport, R. K. & Rogers, C. M. 1963. The effects of environmental restriction upon the chimpanzee's responsiveness to objects. *Journal of Comparative and Physiological Psychology*, **56**, 78–85.

Menzel, E. W., Davenport, R. K. & Rogers, C. M. 1970. The development of tool using in wild-born and restriction-reared chimpanzees. *Folia Primatologica*, **12**, 273–83.

Menzel, E. W., Davenport, R. K. & Rogers, C. M. 1972. Protocultural aspects of chimpanzees' responsiveness to novel objects. *Folia Primatologica*, **17**, 161–70.

Merfield, F. G. & Miller, H. 1956. *Gorillas Were My Neighbours*. London: Companion Book Club.

Merrick, N. J. 1977. Social grooming and play behavior of a captive group of chimpanzees. *Primates*, **18**, 215–24.

Meterological Office. 1975. *Tables of Temperature, Relative Humidity and Precipitation for the World*, part IV, *Africa, the Atlantic Ocean South of 35°N and the Indian Ocean*. London: Her Majesty's Stationery Office.

Milton, K. & Demment, M. 1989. Features of meat digestion by captive chimpanzees (*Pan troglodytes*). *American Journal of Primatology*, **18**, 45–52.

Mitani, M. 1990. A note on the present situation of the primate fauna found from south-eastern Cameroon to northern Congo. *Primates*, **31**, 625–34.

Mitchell, B. L. & Holliday, C. S. 1960. A new primate from Nyasaland. *South African Journal of Science*, **56**, 215–22.

Miyamoto, M. M., Slightom, J. L. & Goodman, M. 1987. Phylogenetic relations of humans and African apes from DNA sequences in the α-globin region. *Science*, **238**, 369.

Montagu, M. F. A. 1968. Brains, genes, culture, immaturity, and gestation. In: *Culture: Man's Adaptive Dimension*, ed. M. F. A. Montagu, pp. 102–13. London: Oxford University Press.
Moore, J. 1985. Chimpanzee survey in Mali, West Africa. *Primate Conservation*, **6**, 59–63.
Moore, J. 1986. Arid country chimpanzees. *Anthroquest*, **36**, 8–10.
Moore, J. H. 1974. The culture concept as ideology. *American Ethnologist*, **1**, 537–49.
Moore, O. K. 1952. Nominal definitions of culture. *Philosophy of Science*, **19**, 245–56.
Moreno-Black, G. 1978. The use of scat samples in primate diet analysis. *Primates*, **19**, 215–22.
Morris, K. & Goodall, J. 1977. Competition for meat between chimpanzees and baboons of the Gombe National Park. *Folia Primatologica*, **28**, 109–21.
Morris, R. & Morris, D. 1966. *Men and Apes*. London: Sphere.
Mottershead, G. S. 1963. Experiences with chimpanzees at liberty on islands. *Zoologische Garten*, **28**, 31–3.
Mundinger, P. C. 1980. Animal cultures and a general theory of cultural evolution. *Ethology and Sociobiology*, **1**, 183–223.
Murdock, G. P. & Provost, C. 1973. Measurement of cultural complexity. *Ethnology*, **12**, 379–92.
Napier, J. R. & Napier, P. H. 1967. *A Handbook of Living Primates*. London: Academic Press.
Nash, V. J. 1982. Tool use by captive chimpanzees at an artificial termite mound. *Zoo Biology*, **1**, 211–21.
Natale, F., Poti, P. & Spinozzi, G. 1988. Development of tool use in a macaque and a gorilla. *Primates*, **29**, 413–16.
Nishida, T. 1968. The social group of wild chimpanzees in the Mahali Mountains. *Primates*, **9**, 167–224.
Nishida, T. 1970. Social behavior and relationships among wild chimpanzees of the Mahali Mountains. *Primates*, **11**, 47–87.
Nishida, T. 1972. Preliminary information of the pygmy chimpanzees (*Pan paniscus*) of the Congo Basin. *Primates*, **13**, 415–25.
Nishida, T. 1973. The ant-gathering behaviour by the use of tools among wild chimpanzees of the Mahali Mountains. *Journal of Human Evolution*, **2**, 357–70.
Nishida, T. 1977. Chimpanzee anting behaviour and its eco-evolutionary significance. In: *Ecology: Course of Physical Anthropology*, vol. 12, ed. H. Watanabe, pp. 55–84. Tokyo: Yuzankaku. [In Japanese.]
Nishida, T. 1979. The social structure of chimpanzees of the Mahale Mountains. In: *The Great Apes*, ed. D. A. Hamburg and E. R. McCown, pp. 72–121. Menlo Park: Benjamin/Cummings.
Nishida, T. 1980*a*. Local differences in responses to water among wild chimpanzees. *Folia Primatologica*, **33**, 189–209.
Nishida, T. 1980*b*. The leaf-clipping display: a newly-discovered expressive gesture in wild chimpanzees. *Journal of Human Evolution*, **9**, 117–28.
Nishida, T. 1983. Alloparental behaviour in wild chimpanzees of the Mahale Mountains, Tanzania. *Folia Primatologica*, **41**, 1–33.
Nishida, T. 1987. Local traditions and cultural transmission. In: *Primate Societies*, ed. B. B. Smuts, D. L. Cheney, R. M. Seyfarth, R. W.

Wrangham and T. T. Struhsaker, pp. 462–74. Chicago: University of Chicago Press.

Nishida, T. 1989. A note on the ecology of the Ugalla area, Tanzania. *Primates*, **30**, 129–38.

Nishida, T. (ed.) 1990. *The Chimpanzees of the Mahale Mountains*. Tokyo: University of Tokyo Press.

Nishida, T. 1992. Meat-sharing as a political strategy of an alpha male chimpanzee? In: *Proceedings of the Symposia of XIIIth Congress of International Primatological Society*, ed. T. Nishida, W. C. McGrew, P. Marler, M. Pickford and F. B. M. de Waal. Tokyo: University of Tokyo Press.

Nishida, T. & Hiraiwa, M. 1982. Natural history of a tool-using behaviour by wild chimpanzees in feeding upon wood-boring ants. *Journal of Human Evolution*, **11**, 73–99.

Nishida, T. & Uehara, S. 1980. Chimpanzees, tools and termites: another example from Tanzania. *Current Anthropology*, **21**, 671–2.

Nishida, T. & Uehara, S. 1983. Natural diet of chimpanzees (*Pan troglodytes schweinfurthii*): long-term record from the Mahale Mountains, Tanzania. *African Study Monographs*, **3**, 109–30.

Nishida, T., Uehara, S. & Ramadhani, N. 1979. Predatory behavior among wild chimpanzees of the Mahale Mountains. *Primates*, **20**, 1–20.

Nishida, T., Wrangham, R. W., Goodall, J. & Uehara, S. 1983. Local differences in plant-feeding habits of chimpanzees between the Mahale Mountains and Gombe National Park. *Journal of Human Evolution*, **12**, 467–80.

Nissen, H. W. 1931. A field study of the chimpanzee. Observations of chimpanzee behavior and environment in western French Guinea. *Comparative Psychology Monographs*, **8**(36): 1–122.

Nissen, H. W. & Crawford, M. P. 1936. A preliminary study of food-sharing behavior in young chimpanzees. *Journal of Comparative Psychology*, **22**, 383–419.

Noble, W. & Davidson, I. 1991. The evolutionary emergence of modern human behaviour: language and its archaeology. *Man*, **26**, 223–53.

Norikoshi, K. 1983. Prevalent phenomenon of predation observed among wild chimpanzees of the Mahale Mountains. *Journal of the Anthropological Society of Nippon*, **91**, 475–9.

Oakley, K. P. 1965. *Man the Tool-Maker*, 5th edn. London: British Museum (Natural History).

O'Connell, J. F., Hawkes, K. & Blurton Jones, N. 1988. Hadza scavenging: implications for Plio-Pleistocene hominid subsistence. *Current Anthropology*, **29**, 356–63.

Oswalt, W. H. 1973. *Habitat and Technology*. New York: Holt, Rinehart and Winston.

Oswalt, W. H. 1976. *An Anthropological Analysis of Food-Getting Technology*. New York: John Wiley.

Oswalt, W. H. n.d. A panspecies analysis of technological forms. Unpublished manuscript.

Palca, J. 1986. Stones turn up. *Nature*, **320**, 3.

Parker, C. E. 1969. Responsiveness, manipulation, and implementation behavior in chimpanzees, gorillas, and orang-utans. In: *Proceedings of the Second International Congress of Primatology, Atlanta, GA, 1968*, vol. 1, *Behavior*, ed. C. R. Carpenter, pp. 160–6. Basel: S. Karger.

Parker, S. & Gibson, K. R. 1979. A developmental model for the evolution of language and intelligence in early hominids. *Behavioral and Brain Sciences*, **2**, 367–408.
Patterson, F. 1986. The mind of the gorilla: conversation and conservation. In: *Primates: The Road to Self-Sustaining Populations*, ed. K. Benirschke, pp. 933–47. New York: Springer.
Peters, C. R. 1982. Electron-optical microscopic study of incipient dental microdamage from experimental seed and bone crushing. *American Journal of Physical Anthropology*, **57**, 283–301.
Peters, C. R. 1987*a*. Nut-like oil seeds: food for monkeys, chimpanzees, humans, and probably ape-men. *American Journal of Physical Anthropology*, **73**, 333–63.
Peters, C. R. 1987*b*. *Ricinodendron rautanenii* (Euphorbiaceae): Zambezian wild food plant for all seasons. *Economic Botany*, **41**, 494–502.
Peters, C. R. & O'Brien, E. M. 1981. The early hominid plant-food niche: insights from an analysis of plant food exploitation by *Homo, Pan*, and *Papio* in eastern and southern Africa. *Current Anthropology*, **22**, 127–40.
Peters, C. R. & O'Brien, E. M. 1982. On early hominid plant-food niches. *Current Anthropology*, **23**, 214–18.
Pettet, A. 1975. Defensive stoning by baboons. *Nature*, **258**, 549.
Pickford, M. 1975. Defensive stoning by baboons. *Nature*, **258**, 549–50.
Pickford, M. 1986. Did *Kenyapithecus* use stones? *Folia Primatologica*, **47**, 1–7.
Plomley, N. J. B. 1966. *Friendly Mission. The Tasmanian Journals and Papers of George Augustus Robinson 1829–1834*. Hobart: Tasmanian Historical Research Association.
Plooij, F. X. 1978*a*. Some basic traits of language in wild chimpanzees? In: *Action, Gesture and Symbol: The Emergence of Language*, ed. A. Lock, pp. 111–31. London: Academic Press.
Plooij, F. X. 1978*b*. Tool-use during chimpanzees' bushpig hunt. *Carnivore*, **1**, 103–6.
Plooij, F. X. 1984. *The Behavioral Development of Free-living Chimpanzee Babies and Infants*. Norwood: Ablex.
Pomeroy, D. E. 1977. The distribution and abundance of large termite mounds in Uganda. *Journal of Applied Ecology*, **14**, 465–75.
Post, D. G., Hausfater, G. & McCuskey, S. A. 1980. Feeding behavior of yellow baboons (*Papio cynocephalus*): relationships to age, gender, and dominance rank. *Folia Primatologica*, **34**, 170–95.
Potts, R. 1984. Hominid hunters? Problems of identifying the earliest hunter. In: *Hominid Evolution and Community Ecology*, ed. R. Foley, pp. 129–66. London: Academic Press.
Potts, R. 1988. *Early Hominid Activities in Olduvai Gorge*. New York: Aldine de Gruyter.
Prasad, K. N. 1982. Was *Ramapithecus* a tool-user? *Journal of Human Evolution*, **11**, 101–4.
Premack, D. 1971. Language in chimpanzee. *Science*, **172**, 808–22.
Premack, D. 1975. Putting a face together. *Science*, **188**, 228–36.
Premack, D., Woodruff, G. & Kennel, K. 1978. Paper-marking test for chimpanzee: simple control for social cues. *Science*, **202**, 903–5.
Rahm, U. 1971. L'emploie d'outils par les chimpanzés de l'ouest de la Côte-d'Ivoire. *Terre et la Vie*, **25**, 506–9.
Redford, K. H. & Dorea, J. G. 1984. The nutritional value of invertebrates

with emphasis on ants and termites as food for mammals. *Journal of Zoology*, **203**, 385–95.

Redshaw, M. 1978. Cognitive development in human and gorilla infants. *Journal of Human Evolution*, **7**, 133–41.

Reynolds, V. 1967. *The Apes*. London: Cassell.

Reynolds, V. 1975. How wild are the Gombe chimpanzees? *Man*, **10**, 123–5.

Reynolds, V. & Reynolds, F. 1965. Chimpanzees of the Budongo Forest. In: *Primate Behavior*, ed. I. DeVore, pp. 368–424. New York: Holt, Rinehart and Winston.

Rhine, R. J., Norton, G. W., Wynn, G. M., Wynn, R. D. & Rhine, H. B. 1986. Insect and meat eating among infant and adult baboons (*Papio cynocephalus*) of Mikumi National Park, Tanzania. *American Journal of Physical Anthropology*, **70**, 105–18.

Rhine, R. J. & Westlund, B. J. 1978. The nature of a primary feeding habit in different age–sex classes of yellow baboons (*Papio cynocephalus*). *Folia Primatologica*, **30**, 64–79.

Rijksen, H. B. 1978. *A Fieldstudy on Sumatran Orang Utans* (Pongo pygmaeus abellii *Lesson 1827*). Wageningen: H. Veenman and Zonen B. V.

Riss, D. C. & Busse, C. D. 1977. Fifty-day observation of a free-ranging adult male chimpanzee. *Folia Primatologica*, **28**, 283–97.

Rodman, P. S. 1977. Feeding behaviour of orang-utans of the Kutai Nature Reserve, East Kalimantan. In: *Primate Ecology*, ed. T. H. Clutton-Brock, pp. 383–413. London: Academic Press.

Rodman, P. S. & McHenry, H. M. 1980. Bioenergetics and the origin of hominid bipedalism. *American Journal of Physical Anthropology*, **52**, 103–6.

Rodriguez, E., Aregullin, M., Nishida, T., Uehara, S., Wrangham, R. W., Abramowski, Z., Finlayson, A. & Towers, G. H. N. 1985. Thiarubrine A, a bioactive constituent of *Aspilia* (Asteraceae) consumed by wild chimpanzees. *Experientia*, **41**, 419–20.

Ron, T. & McGrew, W. C. 1988. Ecological assessment for a chimpanzee rehabilitation project in northern Zambia. *Primate Conservation*, **9**, 37–41.

Rumbaugh, D. M. 1970. Learning skills of anthropoids. In: *Primate Behavior. Developments in Field and Laboratory Research*, ed. L. A. Rosenblum, pp. 1–70. New York: Academic Press.

Rumbaugh, D. M. (ed.) 1977. *Language Learning by a Chimpanzee. The Lana Project*. New York: Academic Press.

Russon, A. E. & Galdikas, B. M. F. 1992. Imitation in ex-captive orangutans (*Pongo pygmaeus*). *Journal of Comparative Psychology* (in press).

Ryan, A. S. & Johanson, D. C. 1989. Anterior dental microwear in *Australopithecus afarensis*: comparisons with human and nonhuman primates. *Journal of Human Evolution*, **18**, 235–68.

Sabater Pí, J. 1974. An elementary industry of the chimpanzees in the Okorobiko Mountains, Rio Muni (Republic of Equatorial Guinea), West Africa. *Primates*, **15**, 351–64.

Sabater Pí, J. 1979. Feeding behaviour and diet of chimpanzees (*Pan troglodytes troglodytes*) in the Okorobiko Mountains of Rio Muni (West Africa). *Zeitschrift für Tierpsychologie*, **50**, 265–81.

Sahlins, M. D. 1959. The social life of monkeys, apes and primitive man. In:

The Evolution of Man's Capacity of Culture, ed. J. N. Spuhler, pp. 54–73. Detroit: Wayne State University Press.

Satterthwait, L. D. 1979. A comparative study of Australian aboriginal food-procurement technologies. PhD thesis, University of California, Los Angeles.

Satterthwait, L. D. 1980. Aboriginal Australia: the simplest technologies? *Archaeology and Physical Anthropology of Oceania*, **15**, 153–6.

Savage, T. S. & Wyman, J. 1844. Observations on the external characters and habits of the *Troglodytes Niger*, Geoff. and on its organization. *Boston Journal of Natural History*, **4**, 362–86.

Savage-Rumbaugh, E. S. 1986. *Ape Language: From Conditioned Response to Symbol*. New York: Columbia University Press.

Savage-Rumbaugh, E. S., Rumbaugh, D. M. & Boysen, S. 1978. Linguistically-mediated tool use and exchange by chimpanzees (*Pan troglodytes*). *Behavioral and Brain Sciences*, **1**, 539–54.

Sayer, J. A. 1977. Conservation of large mammals in the Republic of Mali. *Biological Conservation*, **12**, 245–63.

Schaller, G. B. 1963. *The Mountain Gorilla*. Chicago: University of Chicago Press.

Schaller, G. B. & Lowther, G. R. 1969. The relevance of carnivore behavior to the study of early hominids. *Southwestern Journal of Anthropology*, **25**, 307–41.

Schiller, P. H. 1952. Innate constituents of complex responses in primates. *Psychological Reviews*, **59**, 177–91.

Schneider, A. & Sambou, K. 1982. Prospection botanique dans les Parcs nationaux du Niokolo-Koba et de Basse Casamance. *Mémoires de l'Institut Fondamental d'Afrique Noire*, **92**, 101–22.

Schrire, C. 1980. An inquiry into the evolutionary status and apparent identity of San hunter-gatherers. *Human Ecology*, **8**, 9–32.

Selander, R. K. 1972. Sexual selection and dimorphism in birds. In: *Sexual Selection and the Descent of Man 1871–1971*, ed. B. Campbell, pp. 180–230. Chicago: Aldine.

Sept, J. M. 1986. Plant foods and early hominids at Site FxJj 50, Koobi Fora, Kenya. *Journal of Human Evolution*, **15**, 751–70.

Sept, J. M. 1992. Was there no place like home? A new perspective on early hominid archaeological sites from the mapping of chimpanzee nests. *Current Anthropology*, **33**, 187–207.

Seward, J. H. 1968. Causal factors and processes in the evolution of pre-farming societies. In: *Man the Hunter*, ed. R. B. Lee and I. DeVore, pp. 321–34. Chicago: Aldine-Atherton.

Shipman, P. 1986. Scavenging or hunting in early hominids: theoretical framework and tests. *American Anthropologist*, **88**, 27–43.

Sibley, C. G., Comstock, J. A. & Ahlquist, J. E. 1990. DNA hybridization evidence of hominoid phylogeny: a reanalysis of the data. *Journal of Molecular Evolution*, **30**, 202–36.

Silberbauer, G. B. 1972. The G/wi Bushmen. In: *Hunters and Gatherers Today*, ed. M. G. Bicchieri, pp. 271–326. New York: Holt, Rinehart and Winston.

Silk, J. B. 1978. Patterns of food sharing among mother and infant chimpanzees at Gombe National Park, Tanzania. *Folia Primatologica*, **29**, 129–41.

Silk, J. B. 1979. Feeding, foraging, and food sharing behavior of immature chimpanzees. *Folia Primatologica*, **31**, 123–42.

Simpson, M. J. A. 1973. The social grooming of male chimpanzees. In: *Comparative Ecology and Behaviour of Primates*, ed. R. P. Michael and J. H. Crook, pp. 411–505. London: Academic Press.

Smalley, I. 1979. Moas as rockhounds. *Nature*, **281**, 103–4.

Smith, D. A. 1973. Systematic study of chimpanzee drawing. *Journal of Comparative and Physiological Psychology*, **82**, 406–14.

Snowdon, C. T. 1990. Language capacities of nonhuman animals. *Yearbook of Physical Anthropology*, **33**, 215–43.

Solomon, S., Minnegal, M. & Dwyer, P. 1986. Bower birds, bones and archaeology. *Journal of Archaeological Science*, **13**, 307–18.

Solway, J. S. & Lee, R. B. 1990. Foragers, genuine or spurious? *Current Anthropology*, **31**, 109–46.

Spencer, F., Boaz, N. T., Allen, M. & McGrew, W. C. 1982. Biochemical detection of fecal hematin as a test for meat-eating in the chimpanzee (*Pan troglodytes*). *American Journal of Primatology*, **3**, 327–32.

Speth, J. D. 1987. Early hominid subsistence strategies in seasonal habitats. *Journal of Archaeological Science*, **14**, 13–29.

Speth, J. D. 1989. Early hominid hunting and scavenging: the role of meat as an energy source. *Journal of Human Evolution*, **18**, 329–43.

Spinage, C. A. 1980. Parks and reserves in Congo Brazzaville. *Oryx*, **15**, 292–5.

Stacey, P. B. & Koenig, W. D. 1984. Cooperative breeding in the acorn woodpecker. *Scientific American*, **251**(2), 114–21.

Stahl, A. B. 1984. Hominid dietary selection before fire. *Current Anthropology*, **25**, 151–68.

Standen, V. & Foley, R. A. (eds.) 1989. *Comparative Socioecology*. Oxford: Blackwell Scientific.

Stanford, C. B. & Allen, J. S. 1991. On strategic storytelling: current models of human behavioral evolution. *Current Anthropology*, **32**, 58–61.

Steele, J. 1989. Hominid evolution and primate social cognition. *Journal of Human Evolution*, **18**, 421–32.

Steiner, S. M. 1990. Handedness in chimpanzees. *Friends of Washoe*, **9**, 9–19.

Steklis, H. D. 1990. Chimpanzees in an unusual habitat – the Ishasha River region, Eastern Zaïre. *American Journal of Primatology*, **20**, 235–6.

Steklis, H. D. & Walter, A. 1991. Culture, biology, and human behavior: a mechanistic approach. *Human Nature*, **2**, 137–69.

Stephan, H. & Andy, O. J. 1969. Quantitative comparative neuroanatomy of primates: an attempt at a phylogenetic interpretation. *Annals of the New York Academy of Sciences*, **167**, 370–87.

Stephenson, G. R. 1973. Testing for group specific communication patterns in Japanese macaques. In: *Precultural Primate Behavior*, ed. E. W. Menzel, pp. 51–75. Basel: S. Karger.

Stott, K. & Selsor, C. J. 1959. Chimpanzees in western Uganda. *Oryx*, **5**, 108–15.

Struhsaker, T. T. 1975. *The Red Colobus Monkey*. Chicago: University of Chicago Press.

Struhsaker, T. T. & Hunkeler, P. 1971. Evidence of tool-using by chimpanzees in the Ivory Coast. *Folia Primatologica*, **15**, 212–19.

Suarez, S. D. & Gallup, G. G. 1981. Self-recognition in chimpanzees and orangutans, but not gorillas. *Journal of Human Evolution*, **10**, 175–88.

Sugardjito, J. & Nurhuda, N. 1981. Meat-eating behaviour in wild orang utans, *Pongo pygmaeus*. *Primates*, **22**, 414–16.

Sugiyama, Y. 1968. Social organization of chimpanzees in the Budongo Forest, Uganda. *Primates*, **9**, 225–58.

Sugiyama, Y. 1969. Social behavior of chimpanzees in the Budongo Forest, Uganda. *Primates*, **10**, 197–225.

Sugiyama, Y. 1973. The social structure of wild chimpanzees. A review of field studies. In: *Comparative Ecology and Behaviour of Primates*, ed. R. P. Michael and J. H. Crook, pp. 375–410. London: Academic Press.

Sugiyama, Y. 1981. Observations on the population dynamics and behavior of wild chimpanzees at Bossou, Guinea, in 1979–1980. *Primates*, **22**, 435–44.

Sugiyama, Y. 1984. Population dynamics of wild chimpanzees at Bossou, Guinea between 1976 and 1983. *Primates*, **25**, 391–400.

Sugiyama, Y. 1985. The brush-stick of chimpanzees found in south-west Cameroon and their cultural characteristics. *Primates*, **26**, 361–74.

Sugiyama, Y. 1989. Description of some characteristic behaviours and discussion on their propagation process among chimpanzees of Bossou, Guinea. In: *Behavioral Studies of Wild Chimpanzees at Bossou, Guinea*, ed. Y. Sugiyama, pp. 43–76. Inuyama: Primate Research Institute.

Sugiyama, Y. 1993. Local variation of tool and tool behavior among wild chimpanzee populations. In *The Use of Tools in Primates*, ed. J. Chavaillon. Paris: Fondation Fyssen.

Sugiyama, Y. & Koman, J. 1979. Tool-using and -making behavior in wild chimpanzees at Bossou, Guinea. *Primates*, **20**, 513–24.

Sugiyama, Y. & Koman, J. 1987. A preliminary list of chimpanzees' alimentation at Bossou, Guinea. *Primates*, **28**, 133–47.

Sugiyama, Y., Koman, J. & Bhoye Sow, M. 1988. Ant-catching wands of wild chimpanzees at Bossou, Guinea. *Folia Primatologica*, **51**, 56–60.

Sumita, K., Kitahara-Frisch, J. & Norikoshi, K. 1985. The acquisition of stone-tool use in captive chimpanzees. *Primates*, **26**, 168–81.

Susman, R. L. (ed.) 1984. *The Pygmy Chimpanzee*. New York: Plenum Press.

Susman, R. L. 1988. Hand of *Paranthropus robustus* from Member 1, Swartkrans: fossil evidence for tool behavior. *Science*, **240**, 781–4.

Sussman, C. 1986. Early tools from Olduvai Gorge. *Anthroquest*, **34**, 10–12.

Suzuki, A. 1965. An ecological study of wild Japanese monkeys in snowy areas – focused on their food habits. *Primates*, **6**, 31–72.

Suzuki, A. 1966. On the insect-eating habits among wild chimpanzees living in the savanna woodland of western Tanzania. *Primates*, **7**, 481–7.

Suzuki, A. 1969. An ecological study of chimpanzees in a savanna woodland. *Primates*, **10**, 103–48.

Suzuki, A. 1971. Carnivority and cannibalism observed among forest-living chimpanzees. *Journal of the Anthropological Society of Nippon*, **79**, 30–48.

Suzuki, A. 1975. The origin of hominid hunting: a primatological perspective. In: *Socioecology and Psychology of Primates*, ed. R. H. Tuttle, pp. 259–78. The Hague: Mouton.

Swedlund, A. C. 1974. The use of ecological hypotheses in australopithecine taxonomy. *American Anthropologist*, **76**, 515–29.

Takahata, Y. 1982. Termite-fishing observed in the M group chimpanzees. *Mahale Mountains Chimpanzee Research Project Ecological Report No. 18.*

Takahata, Y., Hasegawa, T. & Nishida, T. 1984. Chimpanzee predation in the Mahale Mountains from August 1979 to May 1982. *International Journal of Primatology*, **5**, 213–33.

Takahata, Y., Hiraiwa-Hasegawa, M., Takasaki, H. & Nyundo, R. 1986. Newly acquired feeding habits among the chimpanzees of the Mahale Mountains National Park, Tanzania. *Human Evolution*, **1**, 277–84.

Takasaki, H. 1983. Mahale chimpanzees taste mangoes – toward acquisition of a new food item? *Primates*, **24**, 273–5.

Takasaki, H. & Hunt, K. 1987. Further medicinal plant consumption in wild chimpanzees? *African Study Monographs*, **8**, 125–8.

Takasaki, H., Nishida, T., Uehara, S., Norikoshi, K., Kawanaka, K., Takahata, Y., Hiraiwa-Hasegawa, M., Hasegawa, T., Hayaki, H., Masui, K. & Huffman, M. A. 1990. Summary of meteorological data at Mahale Research Camps, 1973–1988. In: *The Chimpanzees of the Mahale Mountains*, ed. T. Nishida, pp. 291–300. Tokyo: University of Tokyo Press.

Tanner, N. M. 1981. *On Becoming Human*. Cambridge: Cambridge University Press.

Tanner, N. M. 1987. The chimpanzee model revisited and the gathering hypothesis. In: *The Evolution of Human Behavior: Primate Models*, ed. W. G. Kinzey, pp. 3–27. Albany: State University of New York Press.

Tanner, N. M. & Zihlman, A. 1976. Women in evolution. I. Innovation and selection in human origins. *Signs*, **1**, 585–608.

Tanno, T. 1981. Plant utilization of the Mbuti Pygmies – with special reference to their material culture and use of wild vegetable foods. *African Study Monographs*, **1**, 1–53.

Teaford, M. F. & Walker, A. 1984. Quantitative differences in dental microwear between primate species with different diets and a comment on the presumed diet of *Sivapithecus*. *American Journal of Physical Anthropology*, **64**, 191–200.

Teleki, G. 1973. *The Predatory Behavior of Wild Chimpanzees*. Lewisburg: Bucknell University Press.

Teleki, G. 1974. Chimpanzee subsistence technology: materials and skills. *Journal of Human Evolution*, **3**, 575–94.

Teleki, G. 1975. Primate subsistence patterns: collector-predators and gatherer-hunters. *Journal of Human Evolution*, **4**, 125–84.

Temerlin, M. K. 1975. *Lucy: Growing Up Human*. Palo Alto: Science and Behavior Books.

Terashima, H. 1980. Hunting life of the Bambote: an anthropological study of hunter-gatherers in a wooded savanna. *Senri Ethnological Studies*, **6**, 223–68.

Testart, A. 1988. Some major problems in the social anthropology of hunter-gatherers. *Current Anthropology*, **29**, 1–31.

Tiger, L. & Fox, R. 1971. *The Imperial Animal*. New York: Dell.

Tindale, N. B. 1977. Adaptive significance of the Panara or grass seed culture of Australia. In: *Stone Tools as Cultural Markers: Change, Evolution and Complexity*, ed. R. V. S. Wright, pp. 345–9. Canberra: Australian Institute of Aboriginal Studies.

Tingpalapong, M., Watson, W. T., Whitmire, R. E., Chapple, F. E. & Marshall, J. T. 1981. Reactions of captive gibbons to natural habitat and wild conspecifics after release. *Natural History Bulletin of the Siam Society*, **29**, 31–40.

Tomasello, M. 1990. Cultural transmission in the tool use and communicatory signalling of chimpanzees? In: *'Language' and Intelligence in Monkeys and Apes*, ed. S. T. Parker and K. R. Gibson, pp. 274–311. Cambridge: Cambridge University Press.

Tomasello, M., Davis-Dasilva, M., Camak, L. & Bard, K. 1987. Observational learning of tool-use by young chimpanzees. *Human Evolution*, **2**, 175–83.

Tooby, J. & DeVore, I. 1987. The reconstruction of hominid behavioral evolution through strategic modelling. In: *The Evolution of Human Behavior: Primate Models*, ed. W. G. Kinzey, pp. 183–237. Albany: State University of New York Press.

Torrence, R. 1983. Time budgeting and hunter-gatherer technology. In: *Hunter-Gatherer Economy in Prehistory: A European Perspective*, ed. G. Bailey, pp. 11–22. Cambridge: Cambridge University Press.

Toth, N. 1985*a*. Archaeological evidence for preferential right-handedness in the Lower and Middle Pleistocene, and its possible implications. *Journal of Human Evolution*, **14**, 607–14.

Toth, N. 1985*b*. The Oldowan reassessed: a close look at early stone artifacts. *Journal of Archaeological Science*, **12**, 101–20.

Toth, N. & Schick, K. D. 1986. The first million years: the archaeology of protohuman culture. *Advances in Archaeological Method and Theory*, vol. 9, ed. M. B. Schiffer, pp. 1–96. Orlando: Academic Press.

Trinkaus, E. & Long, J. C. 1990. Species attribution of the Swartkrans Member 1 first metacarpals: SK 84 and SKX 5020. *American Journal of Physical Anthropology*, **83**, 419–24.

Trivers, R. L. 1971. The evolution of reciprocal altruism. *Quarterly Review of Biology*, **46**, 35–57.

Tullot, I. F. 1951. *El Clima de las Posesiones Espanolas del Golfo de Guinea*. Madrid: Instituto de Estudios Africanos.

Tutin, C. E. G. 1979. Mating patterns and reproductive strategies in a community of wild chimpanzees (*Pan troglodytes schweinfurthii*). *Behavioral Ecology and Sociobiology*, **6**, 29–38.

Tutin, C. E. G. & Fernandez, M. 1983. Gorillas feeding on termites in Gabon, West Africa. *Journal of Mammalogy*, **64**, 530–1.

Tutin, C. E. G. & Fernandez, M. 1984. Nationwide census of gorilla (*Gorilla g. gorilla*) and chimpanzee (*Pan t. troglodytes*) populations in Gabon. *American Journal of Primatology*, **6**, 313–36.

Tutin, C. E. G. & Fernandez, M. 1985. Foods consumed by sympatric populations of *Gorilla g. gorilla* and *Pan t. troglodytes* in Gabon: some preliminary data. *International Journal of Primatology*, **6**, 27–43.

Tutin, C. E. G. & Fernandez, M. 1991. Responses of wild chimpanzees and gorillas to the arrival of primatologists: behaviour observed during habituation. In: *Primate Responses to Environmental Change*, ed. H. O. Box, pp. 187–97. London: Chapman and Hall.

Tutin, C. E. G., Fernandez, M., Rogers, M. E., Williamson, E. A. & McGrew, W. C. 1991. Foraging profiles of sympatric lowland gorillas and chimpanzees in the Lope Reserve, Gabon. *Philosophical Transactions of the Royal Society, Series B*, **334**, 19–26.

Tutin, C. E. G., McGrew, W. C. & Baldwin, P. J. 1983. Social organization of savanna-dwelling chimpanzees, *Pan troglodytes verus*, at Mt. Assirik, Senegal. *Primates*, **24**, 154–73.

Tuttle, R. H. 1986. *Apes of the World*. Park Ridge: Noyes.
Tuttle, R. H. 1988. What's new in African paleoanthropology? *Annual Review of Anthropology*, **17**, 391–426.
Ucko, P. J. 1969. Penis sheaths: a comparative study. *Proceedings of the Royal Anthropological Institute of Great Britain and Ireland*, 67pp.
Uehara, S. 1982. Seasonal changes in the techniques employed by wild chimpanzees in the Mahale Mountains, Tanzania, to feed on termites (*Pseudacanthotermes spiniger*). *Folia Primatologica*, **37**, 44–76.
Uehara, S. 1986. Sex and group differences in feeding on animals by wild chimpanzees in the Mahale Mountains National Park, Tanzania. *Primates*, **27**, 1–14.
Uehara, S. 1990. Utilization patterns of a marsh grassland within the tropical rain forest by the bonobos (*Pan paniscus*) of Yalosidi, Republic of Zaïre. *Primates*, **31**, 311–22.
Verschuren, J. 1978. Les grands mammifères du Burundi. *Mammalia*, **42**, 209–24.
Vincent, A. S. 1984. Plant foods in savanna environments: a preliminary report of tubers eaten by the Hadza of northern Tanzania. *World Archaeology*, **17**, 131–48.
Visalberghi, E. 1987. Acquisition of nut-cracking behaviour by 2 capuchin monkeys (*Cebus apella*). *Folia Primatologica*, **49**, 168–81.
Visalberghi, E. 1990. Tool use in *Cebus*. *Folia Primatologica*, **54**, 146–54.
de Waal, F. B. M. 1978. Exploitative and familiarity-dependent support strategies in a colony of semi-free living chimpanzees. *Behaviour*, **66**, 268–312.
de Waal, F. B. M. 1982. *Chimpanzee Politics*. London: Jonathan Cape.
de Waal, F. B. M. 1986. Imaginative bonobo games. *Zoonooz*, **59**(8), 6–10.
de Waal, F. B. M. 1989. Food sharing and reciprocal obligations among chimpanzees. *Journal of Human Evolution*, **18**, 433–59.
Walker, A. & Teaford, M. 1989. Inferences from quantitative analysis of dental microwear. *Folia Primatologica*, **53**, 177–89.
Wallis, J. 1985. Synchrony of estrous swelling in captive group-living chimpanzees (*Pan troglodytes*). *International Journal of Primatology*, **6**, 335–50.
Washburn, S. L. 1972. Human evolution. In: *Evolutionary Biology*, vol. 6, ed. T. Dobzhansky, M. K. Hecht and W. C. Steere, pp. 349–60. New York: Appleton-Century-Crofts.
Washburn, S. L. & Benedict, B. 1979. Non-human primate culture. *Man*, **14**, 163–4.
Washburn, S. L. & DeVore, I. 1961. The social life of baboons. *Scientific American*, **204**, 62–71.
Washburn, S. L. & Lancaster, C. S. 1968. The evolution of hunting. In: *Man the Hunter*, ed. R. B. Lee and I. DeVore, pp. 293–303. Chicago: Aldine-Atherton.
Washio, K., Misawa, S. & Ueda, S. 1989. Individual identification of non-human primates using DNA fingerprinting. *Primates*, **30**, 217–21.
Watts, D. P. 1984. Composition and variability of mountain gorilla diets in the central Virungas. *American Journal of Primatology*, **7**, 323–56.
Watts, D. P. 1989. Ant eating behavior of mountain gorillas. *Primates*, **30**, 121–5.
Weider, D. L. 1980. Behavioristic operationalism and the life-world:

chimpanzees and chimpanzee researchers in face-to-face interaction. *Sociological Enquiry*, **50**, 75–103.

Weiss, G. 1973. A scientific concept of culture. *American Anthropologist*, **75**, 1376–413.

White, F. J. 1992. Activity budgets, feeding behavior and habitat use of pygmy chimpanzees at Lomako, Zaïre. *American Journal of Primatology* **26**, 215–23.

White, F. J. & Wrangham, R. W. 1988. Feeding competition and patch size in the chimpanzee species *Pan paniscus* and *Pan troglodytes*. *Behaviour*, **105**, 148–64.

Whiten, A. 1989. Transmission mechanisms in primate cultural evolution. *Trends in Ecology and Evolution*, **4**, 61–2.

Whitesides, G. H. 1985. Nut cracking by wild chimpanzees in Sierra Leone, West Africa. *Primates*, **26**, 91–94.

Whitten, A. J. 1982. Diet and feeding behaviour of Kloss gibbons on Siberut Island, Indonesia. *Folia Primatologica*, **37**, 177–208.

Wieland, G. R. 1907. Gastroliths. *Science*, **25**, 66–7.

Wiens, J. A. 1977. On competition and variable environments. *American Scientist*, **65**, 590–7.

Wiessner, P. 1981. Measuring the impact of social ties on nutritional status among the !kung San. *Social Science Information*, **20**, 641–78.

Williams, L. 1980. *The Dancing Chimpanzee*. London: Alison and Busby.

Williamson, E. A. 1988. Behavioural ecology of western lowland gorillas in Gabon. PhD thesis, University of Stirling.

Williamson, E. A., Tutin, C. E. G., Rogers, M. E. & Fernandez, M. 1990. Composition of the diet of the lowland gorillas at Lope in Gabon. *American Journal of Primatology*, **21**, 265–77.

Williston, S. W. 1892. Note on the habits of *Ammophila*. *Entomological News*, **3**, 85–6.

Wilmsen, E. N. 1982. Studies in diet, nutrition, and fertility among a group of Kalahari Bushmen in Botswana. *Social Science Information*, **21**, 95–125.

Wolpoff, M. H. 1971. Competitive exclusion among Lower Pleistocene hominids: the single species hypothesis. *Man*, **60**, 601–14.

Wood, R. J. 1984. Spontaneous use of sticks by gorillas at Howletts Zoo Park, England. *International Zoo News*, **31**(3), 13–18.

Woodburn, J. 1968. An introduction to Hadza ecology. In: *Man the Hunter*, ed. R. B. Lee and I. DeVore, pp. 49–55. Chicago: Aldine-Atherton.

Woodburn, J. 1970. *Hunters and Gatherers. The Material Culture of the Nomadic Hadza*. London: British Museum.

Woodruff, G. & Premack, D. 1981. Primitive mathematical concepts in the chimpanzee: proportionality and numerosity. *Nature*, **293**, 568–70.

Wrangham, R. W. 1974. Artificial feeding of chimpanzees and baboons in their natural habitat. *Animal Behaviour*, **22**, 83–93.

Wrangham, R. W. 1975. The behavioural ecology of chimpanzees in Gombe National Park, Tanzania. PhD thesis, University of Cambridge.

Wrangham, R. W. 1977. Feeding behaviour of chimpanzees in Gombe National Park, Tanzania. In: *Primate Ecology*, ed. T. H. Clutton-Brock, pp. 503–38. London: Academic Press.

Wrangham, R. W. 1980*a*. Bipedal locomotion as a feeding adaptation in gelada baboons, and its implications for hominid evolution. *Journal of Human Evolution*, **9**, 329–31.

Wrangham, R. W. 1980*b*. Leaf-grooming by chimpanzees: a preliminary analysis. Unpublished manuscript.

Wrangham, R. W. 1986. Ecology and social relationships in two species of chimpanzees. In: *Ecological Aspects of Social Evolution*, ed. D. I. Rubenstein and R. W. Wrangham, pp. 352–78. Princeton: Princeton University Press.

Wrangham, R. W. 1987. The significance of African apes for reconstructing human evolution. In: *The Evolution of Human Behavior: Primate Models*, ed. W. G. Kinzey, pp. 28–47. Albany: State University of New York Press.

Wrangham, R. W., Clark, A. P. & Isabirye-Basuta, G. 1992. Female social relationships and social organization of Kibale Forest chimpanzees. In: *Human Origins*, ed. T. Nishida. Tokyo: Tokyo University Press.

Wrangham, R. W. & Nishida, T. 1983. *Aspilia* spp. leaves: a puzzle in the feeding behavior of wild chimpanzees. *Primates*, **24**, 276–82.

Wrangham, R. W. & van Zinnicq Bergmann Riss, E. 1990. Rates of predation on mammals by Gombe chimpanzees, 1972–1975. *Primates*, **31**, 157–70.

Wright, R. V. A. 1972. Imitative learning of a flaked stone technology – the case of an orangutan. *Mankind*, **8**, 296–306.

Wynn, T. 1981. The intelligence of Oldowan hominids. *Journal of Human Evolution*, **10**, 529–41.

Wynn, T. 1988. Tools and the evolution of human intelligence. In: *Machiavellian Intelligence*, ed. R. Byrne and A. Whiten, pp. 271–84. Oxford: Clarendon Press.

Wynn, T. 1989. *The Evolution of Spatial Competence*. Urbana: University of Illinois Press.

Wynn, T. 1992. Layers of thinking in tool behavior. In: *Tools, Language and Intelligence: Evolutionary Implications*, ed. T. Ingold and K. R. Gibson. Cambridge: Cambridge University Press. (In press.)

Wynn, T. & McGrew, W. C. 1989. An ape's view of the Oldowan. *Man*, **24**, 383–98.

Wynn, T. & Tierson, F. 1990. Regional comparisons of the shapes of later Acheulean handaxes. *American Anthropologist*, **92**, 73–84.

Yamagiwa, J., Mwanza, N., Yumoto, T. & Marushashi, T. 1991. Ant eating by eastern lowland gorillas. *Primates*, **32**, 247–53.

Yamagiwa, J., Yumoto, T., Ndunda, M. & Maruhashi, T. 1988. Evidence of tool-use by chimpanzees (*Pan troglodytes schweinfurthii*) for digging a bee nest in the Kahuzi-Biega National Park, Zaïre. *Primates*, **29**, 405–11.

Yellen, J. E. & Lee, R. B. 1976. The Dobe/Du/da environment: background to a hunting and gathering way of life. In: *Kalahari Hunter-Gatherers*, ed. R. B. Lee and I. DeVore, pp. 27–46. Cambridge, Mass.: Harvard University Press.

Yerkes, R. M. 1927. The mind of a gorilla: I and II. *Genetic Psychology Monographs*, **2**, 1–193, 337–551.

Yerkes, R. M. 1933. Genetic aspects of grooming, a socially important primate behavior pattern. *Journal of Social Psychology*, **4**, 3–23.

Yerkes, R. M. & Yerkes, A. W. 1929. *The Great Apes*. New Haven: Yale University Press.

Yunis, J. J. & Prakash, O. 1982. The origin of man: a chromosomal pictorial legacy. *Science*, **215**, 1525–30.

Zeven, A. C. 1972. The partial and complete domestication of the oil palm (*Elaeis guineensis*). *Economic Botany*, **26**, 274–9.

Zihlman, A. L. 1978. Women and evolution. II. Subsistence and social organization among early hominids. *Signs*, **4**, 4–20.

Zihlman, A. L. 1981. Women as shapers of the human adaptation. In: *Woman the Gatherer*, ed. F. Dahlberg, pp. 75–120. New Haven: Yale University Press.

Zihlman, A. L., Cronin, J. E., Cramer, D. L. & Sarich, V. M. 1978. Pygmy chimpanzee as a possible prototype for the common ancestor of humans, chimpanzees and gorillas. *Nature*, **275**, 744–6.

Author index

Subject index